DR MATT'S GUIDE TO LIFE IN SPACE

DR MATT'S GUIDE TO **LIFE IN SPACE**

MATT AGNEW

ALLEN&UNWIN
SYDNEY • MELBOURNE • AUCKLAND • LONDON

PART 3: WHERE ARE THE ALIENS?

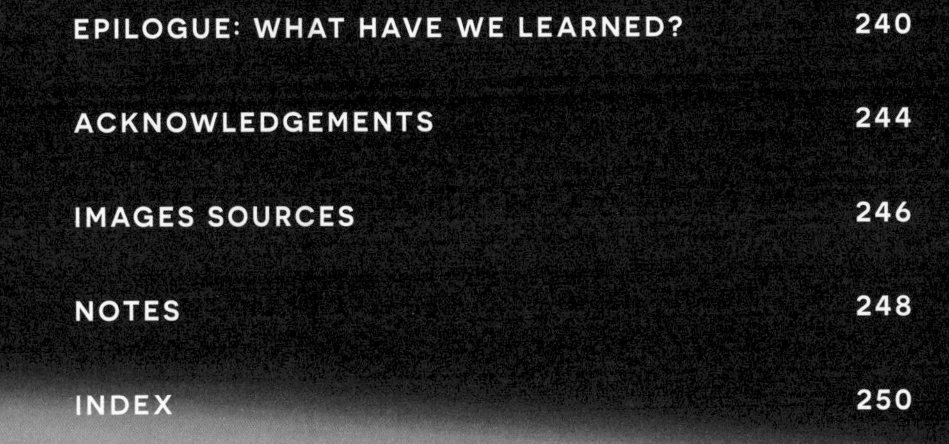

HOW DID WE GET HERE?

We sure didn't muck around with jumping straight into the big questions, did we? *How did we get here?* Not only is it not a gentle question to ease us into things, but it's also one that, well, doesn't have an answer. But it's the perfect way for us to contextualise what it is we want to achieve here with this book. What is it that we want to know? Why did you pick up this book in the first place?

As always, let's start with the first things first. The Big Bang is our best explanation for how the Universe began, and how everything was created. This includes the most fundamental of all the elements: hydrogen. You probably don't stop and think about hydrogen very often, but there is something astonishing, breathtaking, mind-boggling and mind-blowing about hydrogen, and it's this: given enough time, hydrogen eventually starts to ask: 'How did we get here?'

That's right. In some way, *you* exist because of that hydrogen made in the early stages of the Universe. In the last 13.8 billion years, hydrogen has had a pretty wild journey. It has collided and fused into other elements in the core of stars, and it has been violently expelled in the explosions of dying stars. Forming and re-forming, fusing and exploding, and reacting and interacting with other elements. Hydrogen forms stars, heavier elements, planets, bodies of water, simple life forms, complex life forms, language, art, science, philosophy and, eventually, begins to ponder to itself: *How did we get here?*

As Carl Sagan famously said, 'We are a way for the cosmos to know itself.' That is the context to all this. The cosmos is able to learn about itself through us, and an understanding of us inadvertently means we require an understanding of our origins. Where we—life—all began, and why.

Answering these questions is a lofty, if not insurmountable, goal for a book. People have spent years, decades, even their whole lives searching for the answers. In fact, this search is probably not constrained to the length of a single life, but to our whole collective lives as humans. As long as humans have had the cognitive abilities to think and imagine beyond our immediate needs, we have sought to understand why and how we got here. So, while we won't be diving into such profoundly existential questions of our origins entirely within these pages, this book will, however, tantalise your scientific tastebuds, stoke the flame of inquisitiveness and ignite your curiosity. Science continues to move forward and progress because of the collective efforts of many. It is our—my and your—desire to learn that keeps us all asking questions and seeking to understand the world around us. This book covers many topics; I hope it answers some questions you have (and some you might not know you have) and prompts you to ask some new ones.

We'll navigate this book in three parts. And begin with what we know in Part 1: What makes the Earth special? We'll discuss how the Earth came to be, why Earth is such a special place for life, and the different aspects of life—how it got started and evolved to what we see today.

We'll then jump to what we're beginning to know in Part 2: Where else is special? Beyond Earth, are there other places for life in the Solar System? What about beyond that, in other star systems? How do we even begin to look and explore that far away?

FUN BUBBLE!

Welcome to the Fun Bubble! You'll see these bubbles of fun popping up from time to time. Throughout this book, we'll sometimes touch on subjects that have a really cool fact or just warrant a bit of a tangent to discuss them further. While these are all super-interesting in their own right, I want to make this a relaxing scientific journey that doesn't jump around too much. As such, you can read these tasty morsels of knowledge at your leisure.

In Part 3, we can really let the imagination run wild. We'll look at what we don't know: Where are the aliens? How common are they? Have we spotted any? If not, why not? It's safe to say, we've got a lot of ground to cover. So sit back, relax and enjoy the ride. It's going to be a blast. So, let's get into it.

PART ONE

WHAT MAKES EARTH SPECIAL?

The Earth is a pretty special place. But just how special is it? To help answer that, let's take a step back, and say that *the Universe* is a pretty special place. To understand just how special the Universe is, let's put things into context a little. As we noted earlier, the Big Bang is our best explanation for how the Universe began, and how everything was created, including and especially hydrogen (the building block of all other elements). As we highlighted, just as important as the creation of hydrogen is the multi-billion-year journey it proceeds along, which ultimately leads it to end up in such an exotic arrangement of matter that is you and me. The calcium in our bones, the iron in our blood, the carbon that makes up our organic material, the water that courses through our body, these complex elements all have, to varying degrees, the same parent element: hydrogen. And so while somewhat amusing to say, 'given enough time, hydrogen eventually starts to ponder its existence' it is in essence true. Without hydrogen we wouldn't exist, the Earth wouldn't exist, and we wouldn't exist in the way we do today capable of asking, hypothesising and understanding the world around us.

That is the context to all this. The cosmos is able to learn about itself, right here, on Earth. And *that* is how special the Earth is.

So with that in mind, let's dive in. Let's start to really try and wrap our collective heads around why the Earth is special, as that is the first stepping stone in understanding why we are special. What is it exactly that has helped the Earth become the cradle from which myriad life forms have sprung forth, from which intelligent life has emerged and risen to become the custodians of the Earth, and from which civilisation has evolved and thrived to the point of seeking to understand our place in the Universe; how we belong; and how special the Earth is?

CHAPTER 1

WATER

Where else to kick off our discussion about life than with the *elixir of life* itself: water?

Water is fundamentally linked to our existence but we don't often stop to ponder: why is that so? What is it about water that makes it so special to us? Actually, that's a thought! Is it special to *just* us? Or is it special to *all* forms of life? Well, let's dive in (pun *very* intended). Let's break down our all-encompassing quandary about water into some smaller, more easily digestible questions. What's the origin of water? What properties of water make it special? And how do these properties make it integral to the beginning, and continued survival, of life?

To understand the origin of water, first, we need to understand what water is *made of*. What are the building blocks of water? Chemically speaking, water consists of two hydrogen atoms and one oxygen atom, which you may have seen expressed as H_2O. To understand the origin of water, therefore, we must first understand the origins of hydrogen and oxygen. Hydrogen is easy. Basically, without opening an enormous Pandora's box of discussion about the origin of the Universe, it was made during the Big Bang (as we briefly mentioned previously).

THE BIG BANG

BANG!

There it was, the origin of the Universe. Well, we think that was it. It's hard to piece together something that happened 13.8 billion years ago accurately but, as a *hypothesis*, the Big Bang helps to explain several observed phenomena and fits with our understanding of physics as we know it. (That's the basis of science, really; you come up with an idea or hypothesis to explain something, then you carry out tests or *experiments* to reject or support your hypothesis.) In the early stages of the Universe, there was a lot going on really quickly—far too much to describe here—so we're going to fast forward a bit. Not too far though, only a second after the Big Bang, when the Universe has cooled enough that subatomic particles, specifically *protons* and *electrons*, start to form.

THINGS THAT GO BANG!

Whole books have been written on the Big Bang, so we can't really go into huge detail in just one Fun Bubble. Instead, let's take a look at a high-level timeline of the Universe.

Loosely, there are five stages:

1. The very early Universe spans only a fraction of a second. And we believe some bizarre and exotic physics existed. The phenomenon known as *Cosmic Inflation* took place here.
2. The early Universe (spanning about 370,000 years) is where things started to cool down and stuff (that is, matter) started to form. Very simple, but very important, stuff.
3. The Universe then entered the dark ages (spanning 1 billion years). We call this state the 'dark ages' because no stars had formed yet, so there were no sources of light. As such, it was towards the end of this era that large structures such as stars (with planets) and galaxies formed.
4. The Universe as it exists today. After the dark ages, the Universe has evolved slowly and at smaller scales,

and not so much at the largest scales. This means that—as a whole—it has looked pretty similar for the last 12.8 billion years.

5 Obviously, the future of the Universe is unknown. We have some hypotheses about the ultimate fate of the Universe, such as the Big Crunch, the Big Bounce and the Big Freeze, but that discussion is for another book!

Subatomic particles are the *building blocks of the building blocks*. If hydrogen is your cake, then protons and electrons are the flour and eggs. Within a second of the Big Bang, our flour and eggs are there, floating about, but they haven't been combined into cake yet.

Actually, we won't have cake for quite a while. About 370,000 years later, our ingredients (read: subatomic particles) finally combine and our cakes (read: hydrogen atoms) form. After those 370,000 years pass, well, there it is. Hydrogen! Didn't I say hydrogen is easy? We've now got half of our ingredients for water.

What about oxygen then? Well, oxygen is a little trickier. Anything other than hydrogen is formed by a funky process called *nucleosynthesis*. This process essentially means combining simpler elements (such as hydrogen) into heavier, more complex elements. Immediately after the Big Bang, nucleosynthesis took place to form helium (He), a little bit of lithium (Li) *and that's it*. So, the Universe goes *BANG!* We've got loads of hydrogen knocking about, plenty of helium and some lithium, but nothing else. What gives? Fortunately, the Big Bang isn't the only place where nucleosynthesis occurs.

There you are, reading this book (and what a book, right?) on a summer's day. You're outside, basking in the Sun's warmth, wondering what the answer to this chemical conundrum is. And funnily enough, that very Sun you're basking in is the answer to this riddle. Stars, like our Sun, are not just the purveyors of beautiful sunny days, they are pivotal to the existence of the Universe as we know it. Nucleosynthesis takes place deep inside the cores of stars. Stars emit light and heat generated by nuclear fusion, a process in which the star slams lighter elements together to form heavier elements. While the Big Bang got us started with the three lightest elements (hydrogen, helium and lithium), the stars carry us the rest of the way, slamming those lighter elements together into heavier and heavier elements, eventually producing exactly what we're looking for: oxygen. This oxygen is trapped deep inside the star, where the nuclear fusion process is going on. When the star eventually dies, it may explode in an enormous and spectacular explosion called a supernova, spewing all these heavier elements out into space.

SHARING IS CARING

Alright, now we're cooking with gas! We've got the building blocks of water, but we need to figure out how to combine them. How do we put these two separate elements together? The answer comes from a fundamental, but tremendously important, property of nature: nature has preferences. There's more nuance to it than that (far too much to get into here), but the essence is that nature likes to be in particular states, or forms. It likes things to be balanced, stable and in equilibrium. One such preference is for elements to have a full outer shell of electrons, but what does that mean?

From our cake analogy, you'll remember that electrons were our eggs. Let's stick with electrons being eggs again for another analogy, but modify the rest of it a little bit to suit our needs. Through rigorous experimentation, we've found that each element has a number of cartons (or electron shells) and we can fit a certain number of eggs (electrons) in each carton. Now, we're making a few simplifications here because molecular physics gets complicated very quickly, but typically it is only the cartons of eggs that are not full that we are interested in. To make things more confusing, these are weird egg cartons, as they come in different sizes and so can fit different numbers of eggs. Hydrogen has a carton that can fit two eggs, but only has one egg in it. Oxygen, on the other hand, has a carton that can fit eight eggs, but only has six eggs in it. Have a look at Figure 1.1 to see what we're talking about.

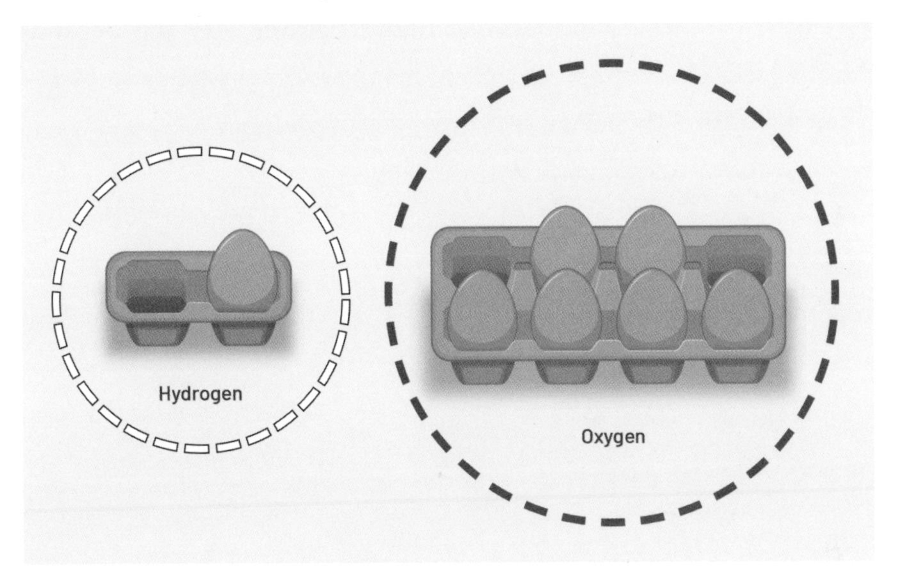

Figure 1.1: Hydrogen (in white), showing one egg in its outer carton (that can hold two eggs), and oxygen (in red), showing six eggs in its outer carton (that can hold eight eggs).

Both these elements—hydrogen and oxygen—really want their cartons to be full.

Nature solves this in a way we were all taught in school: *sharing is caring*. What happens is that oxygen shares one of the eggs in its carton with the empty spot in the hydrogen carton, and another one of its eggs with the other empty spot in a second hydrogen carton. This means the hydrogens now have full two-egg cartons (although that second egg in each is only really half theirs, they need to stay close to oxygen to share it). Similarly, both hydrogens are happy to share their original eggs with the empty spots in oxygen's carton. This means oxygen now fills its eight-egg carton (with four of these original eggs being only in oxygen's carton; two being shared with hydrogen's empty carton spots; and two being received from hydrogen sharing its egg—so all this sharing means they need to stay close to each other), a bit like what is shown in Figure 1.2. This sharing of eggs, or electrons, is known as a *covalent bond*. Nature does this because nature likes elements to have full cartons of eggs. They're happy. They're bonded. By having two hydrogens bonded with one oxygen, we now have exactly what we were looking for: H_2O. Water. When hydrogen from the Big Bang and oxygen from a star meet in the cold vastness of space, they share their eggs to fill their cartons and bond together, forming water.

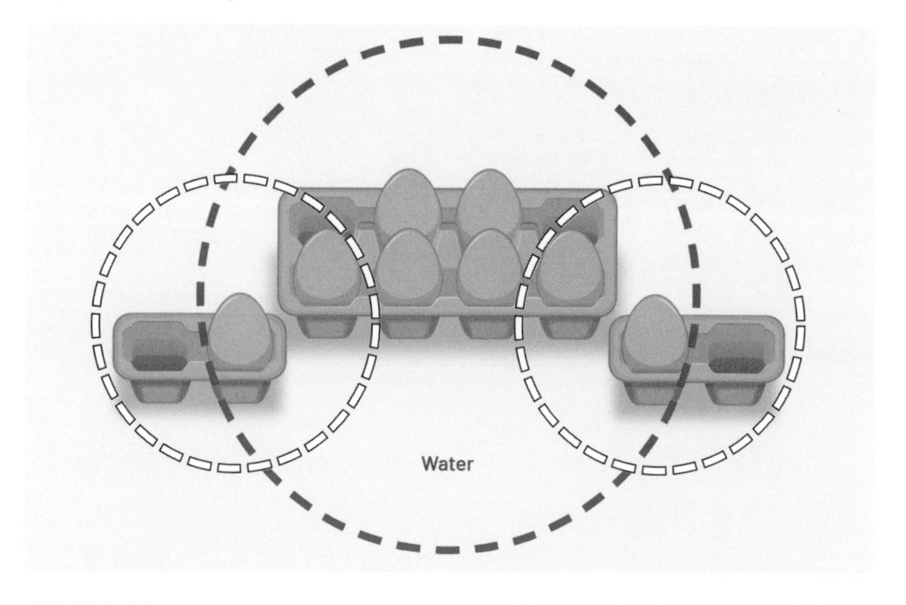

Figure 1.2: Our elements are now sharing eggs (electrons). Both hydrogen atoms (in white) now have two eggs and can fill their cartons, and oxygen (in red) now has eight eggs and can fill its carton. The elements are bonded.

HOW DID WATER GET HERE?

Now that we've got water, we need to understand how it ended up on Earth. The formation of the Solar System is a complex process, which we still don't entirely understand (or even agree on!). While we won't discuss the intricacies of it here, these are the fundamentals of what happens:

1. Lots of gas and dust were floating around in space (created by the Big Bang or older stars that had exploded). This mixture slowly started collapsing in on itself from gravity, slowly accumulating into a larger and larger blob of stuff.

2. Eventually, so much stuff gathered, and it squeezed together so tightly that nuclear fusion (that whole nucleosynthesis process we talked about of slamming elements together) began and the Sun formed.

3. Not all the gas and dust were used to form the Sun; some remained in an enormous disc of gas, dust and ice around the Sun.

4. The planets of the Solar System, including Earth, formed from this disc. In simple terms, some of the dust in the disc collided and stuck together, and then more and more collided with that blob and stuck together, too. This blob of dust kept growing and growing, as more dust collided and stuck to it until—with enough dust—it was the size of a rock, then a boulder, a mountain and, finally, a planet.

If the Earth formed by lots of dust joining together, then where was all the water? Imagine you're sitting around a really hot campfire. It's so hot that it boils any water nearby. If you had a cup of water right next to the campfire, the water boils—it exists only as a gas. You take a step back and grab another cup of water. This one boils, too. As you step further from the campfire, the cup of water gets cooler and cooler until you reach a distance at which the water does not boil—

it remains as a liquid. When discussing planet formation, this is what we call the *ice line* or *snow line*. In this scenario, water near our fire (the Sun) exists as a gas, while water further away exists as a liquid (or in the case of space, ice).

Earth formed in the inner part of the Solar System, where water existed as a gas. As Earth formed into a big ball of rock, plenty of gaseous water was caught between the dust grains and embedded within the ball so, when Earth formed, it was already quite 'wet'. But this was not the only source of water. Far away from the Sun, the water is icy; this is where comets form. Sometimes—in the 4.5 billion years that Earth has existed—one of these icy comets collided with Earth. When this happened, it delivered all that icy water to Earth. These are the two ways in which Earth became the ocean-covered, blue marble we know today: some water was trapped as gas when Earth formed, and other water collided into Earth later, from far outside the Solar System.

Now we have an understanding of not just the origin of water on Earth, but the origin of water *in the Universe!* Let's start going through the properties of water. What is it about water that makes it special? What is it that has made water important to life here on Earth? Several properties of water are critical to making it profoundly special to life as we know it. We'll touch on some of those here, and expand on them further in later chapters.

PROPERTIES OF WATER

The first property of water I want to discuss is its polarity. *Polarity* means that water molecules have *poles* but, rather than being magnetic poles like the Earth has, they are electric poles. To better understand these electric poles, we'll need to understand the geometry, or shape, of a water molecule. Water molecules aren't straight lines. They're sort of like little arrows or corners. Figure 1.3 shows what a water molecule looks like (I've let go of the cartons of eggs here, so things don't get too messy). Here, we can see the big oxygen atom's nucleus in red, the smaller hydrogen nuclei in white, and the electrons as small black dots. The grey bars between the nuclei indicate the covalent bonds holding the elements together (the egg-sharing in our egg-carton analogy).

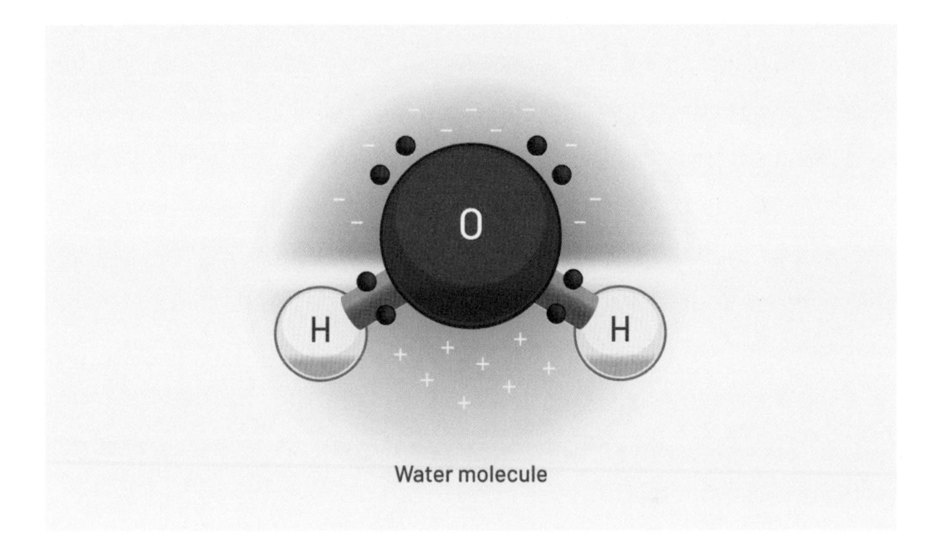

Figure 1.3: A water molecule. The red circle represents the oxygen atom and the white circles represent the hydrogen atoms. The black dots represent electrons, and the grey bars indicate the covalent bond holding the molecule together. Around the molecule, the red shading indicates a slight negative charge, while the blue shading indicates a slight positive charge.

You will notice that the four electrons that oxygen *isn't* sharing all cluster on one side of the water molecule. This is because oxygen has a higher *electronegativity* than hydrogen; it attracts electrons more strongly. Because those four electrons all have a negative charge, there is an overall (albeit very small) negative charge on that side of the water molecule. This is shown with the soft red shading.

In contrast, because hydrogen only has one electron, which it's sharing with oxygen to form the covalent bond (and which is pulled nearer to oxygen because of oxygen's higher electronegativity), there are no electrons on the opposite side of the water molecule. Compared to the oxygen side of the molecule, there is an overall (albeit small) positive charge. We show this with the soft blue shading.

What does all this mean? It means that our water molecule, although quite simple, has positively and negatively charged sides. It has poles! It has an electric direction. This may not seem exciting, but several properties of water hinge on the fact that water has poles. And because it has two poles, water has *dipoles*.

Why do the dipoles of water make it so special? So far, we have discussed *intramolecular bonds*; that is, bonds that occur within a molecule (our cartons sharing eggs). As the name suggests, there are also *intermolecular bonds*; that is, bonds that occur between molecules. The bonding between one water molecule and another happens because of the dipoles. Just like moving one bar magnet towards another will cause the north and south ends to face each other and be held together (or bonded), the slightly negative oxygen side of a water molecule will also try to connect or bond with the slightly positive hydrogen side of another water molecule.

This type of bonding is referred to as *hydrogen bonding* (because, funnily enough, a hydrogen atom is involved!). As far as bonds *between* molecules go, it is quite strong. Not as strong as the egg-sharing (covalent) bond, taking place between the hydrogen and oxygen in our molecule, but it's still reasonably strong. This bond strength gives water a high *specific heat capacity* (meaning it takes a lot of energy to increase the temperature of water), a high *enthalpy of fusion* (the amount of energy it takes to turn ice into water at 0°C), and a high *enthalpy of vaporisation* (the amount of energy it takes to turn water into steam at 100°C). These three properties are important to maintaining a stable atmosphere, which is capable of sustaining life. We'll discuss this further in Chapter 2.

Another key property of water created by the dipoles and intermolecular hydrogen bonds is the density of water. Specifically, solid water is lighter than liquid water. You've probably never thought about it, but it's super-weird.

Typically, the gas form of a molecule is the lightest, then the liquid form and finally the solid form. Why then are both gas and solid water (ice) lighter than liquid water? Ice is the odd one out; it's because the hydrogen bonds between molecules create a sort of crystal lattice or grid. This stops the molecules from getting closer and closer together, and keeps big spaces between molecules. These spaces make ice float and, while this may not sound like a big deal, it has profound implications for life. We'll discuss this further in Chapter 6.

Another special (and slightly strange) property of water I want to discuss is called its *triple point*. These are the conditions (temperature and pressure) in which water can exist as a solid, liquid or gas. It's weird to think that such conditions exist, but it's even weirder that those conditions are remarkably close to the average conditions here on Earth. That's right, we don't need some wacky, futuristic, hypercharged reactor to make it happen. No, under fairly normal conditions, water can exist in all three states. Right now, you could be sitting there with a glass of liquid water next to you, with cubes of solid ice bobbing in it, and the kettle boiling with gaseous water in the kitchen. Water in all three states of matter can exist right next to each other. Think about the water cycle, shown in Figure 1.4. Water can exist as a liquid, be evaporated by the sun and hover in the atmosphere as a gas, only to be condensed back to water where it can rain down or—if the temperature is cold enough—freeze and snow down to Earth. Water cycles every day from one state to another with ease. To me, that's weird. And this makes water really special and important for processes such as plate tectonics, which we'll discuss further in Chapter 3.

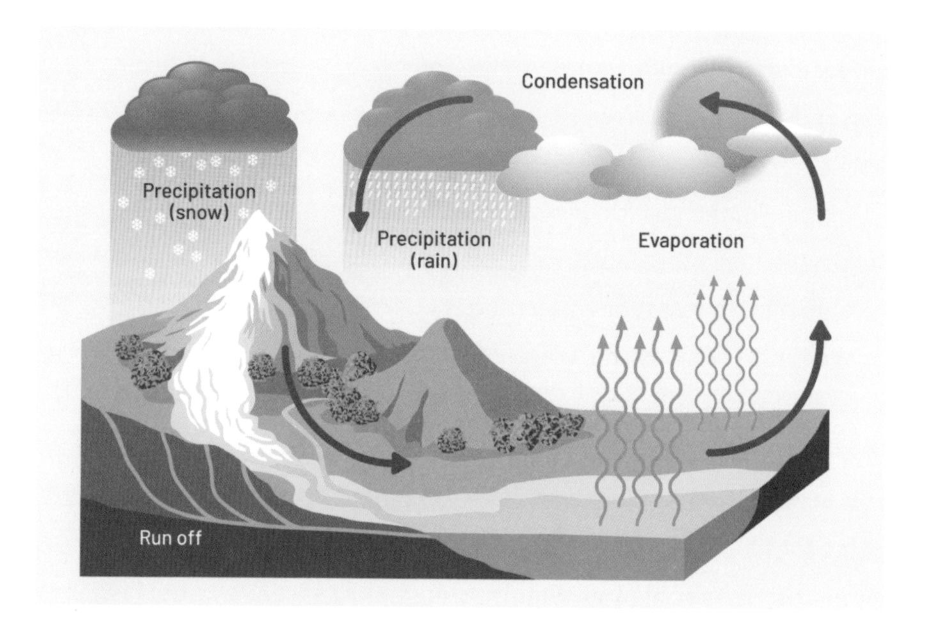

Figure 1.4: The water cycle. Water can exist as a liquid on Earth's surface. Heat from the Sun causes the water to evaporate as a gas. It then rises high into the atmosphere. From here, the water gas can condense back to a liquid and rain back down to Earth or if the conditions are cold enough, that liquid will freeze to a solid and snow back down to Earth. The snow will then melt as it heats, turn back into a liquid and the whole process—or cycle—can continue.

Finally, I want to touch on one last aspect of water that makes it particularly important for life (and that again hinges on its dipoles): its ability to act as a solvent. A solvent is any substance in which something can dissolve. Maybe you've stirred some sugar into your tea (the kettle was boiling, after all), or copped a mouthful of salt water after being dumped by a wave at the beach. These are *solutions* of sugar dissolved in water (the tea), or salt dissolved in water (the salty water of the ocean). Because of its dipoles, water is great at dissolving things that are polar, or have a charge. Because of this property of water being *really* good at dissolving things; water is sometimes referred to as the 'universal solvent', as it can typically dissolve more substances than any other liquid. But why should it matter that water is so good at dissolving stuff?

Water's ability to dissolve so many things matters for one very important reason, which was summed up succinctly by the 1937 winner of the Nobel Prize in Physiology and Medicine, Albert Szent-Györgyi: 'Water is life's matter and matrix, mother and medium.'

You see, because water can dissolve so many things, it can act as a medium for things to move through. Not just living things, either. It's not only a highway for microbial life, but also for the fundamental ingredients of life. Life-building proteins. Information-bearing nucleic acids. Spectacularly rich biological molecules and nutrients all flow through this matrix of life, colliding, bouncing around, forming and re-forming in a great aquatic ballet. This extraordinary dance is only possible because water has that seemingly simple ability to dissolve stuff. When you stirred sugar into your tea this morning, did you think for a moment that you were utilising water's arguably greatest power? We'll go into much more detail about the watery dance of life in a later chapter, but for now let's look deeper into transporting things about the body.

We've just run through a lot of detail about properties of water, so let's quickly summarise them. Water has:

» high specific heat capacity (energy needed to change the temperature of water)
» high enthalpies of fusion and vaporisation (energy needed to change the state of water)
» low density of ice (it floats)
» dipoles (leading it to be called the 'universal solvent').

All these properties contribute to water being a truly special and remarkable substance when it comes to supporting life.

Water's high specific heat capacity (due to the strong hydrogen bonds *between* water molecules) means that our oceans can regulate temperatures on Earth effectively, buffering huge temperature fluctuations. Water's high enthalpies of fusion and vaporisation compound this effect, helping to maintain a stable atmosphere on Earth.

The low density of ice provides an important environment for life in bodies of water. Imagine a lake in a cold climate. If colder water was denser than warmer water (as is typical for a lot of substances), then the coldest water would be on the bottom of the lake. As it chilled further, the lake would freeze from the bottom up, eventually causing the entire lake to be frozen solid and killing any living thing within it. Not a pleasant scenario, you'd agree! Even worse, because of the large amount of energy required to melt the ice (the enthalpy of fusion), the lake may never thaw out. Fear not, though; because of water's odd density, ice will freeze on the top of the lake, and the densest water (which is about 4°C) will actually be beneath this ice. Now, we have liquid water underneath the ice, in which animals can live, and the layer of ice acts like a greenhouse or swimming pool cover, helping to trap heat in the water. Who would have thought that the ice cubes floating in your beverage of choice are linked to a property that ensures life can exist in large bodies of water?

Our bodies are made up of so much water (on average, around 60 per cent!) because the dipoles of water allow it to be the universal solvent. Because water can dissolve nutrients, when it flows, it brings these nutrients along for the ride. This is exactly what happens in our bodies. Oxygen and other nutrients hitch a ride on the water superhighway flowing around our bodies, to where they need to be—supplying our muscles and organs with the vital ingredients to continue functioning. Similarly, waste is removed by water, flowing through our bodies until eventually being expelled. (I'll refrain from injecting toilet humour here; that's just not something I would do-do.)

THE SOUNDS OF CELSIUS

Have you ever noticed that hot and cold water sound different? In case you hadn't, go boil a kettle and fill a bottle with cold water. Close your eyes and have a friend pour a cup from the kettle, and another cup from the bottle. You'll surprise yourself when you can pick which is which! Why is that the case? It's due to a property of fluids called viscosity, which is sort of how 'sticky' a fluid is. If you pour water from a bottle, it'll flow quickly, because it has a low viscosity. If you pour honey out of a bottle, it'll . . . well, it sort of flows, but really slowly because it has a high viscosity. Detergent is somewhere in-between. The viscosity of something changes with temperature—the hotter something is, the more it'll flow (it's runnier, right?). What's happening with our singing water then? Cold water has a higher viscosity; it's a little more 'sticky'. When it hits the cup, it bounces around, making a low-frequency sound. In contrast, hot water has a lower viscosity and is a little less 'sticky'. When it hits the cup, it bounces around a little more, making a high-frequency sound. These differences are subtle, but our keen sense of hearing is clever enough to spot (or hear!) the difference; hence, we have our hot soprano water and our cold baritone water.

In this chapter, we've discussed the origin of water, how it got to Earth (some was trapped when Earth formed, some arrived later from comets). We've also looked at the properties of water that make it so special, not just to us but to Earth; how water has made our little blue marble in space the flourishing greenhouse of creation for us and so many other life forms. A stable atmosphere is also important to life, so let's dive into that topic a little more.

CHAPTER **2**

ATMOSPHERES

Breathe in.

Breathe out.

Inhale.

Exhale.

Feels good, doesn't it? This whole breathing thing is a bit of alright! Breathing is a form of *respiration* which, physiologically speaking, is the process of moving oxygen from outside to inside our bodies, while also moving carbon dioxide from inside to outside our bodies. For us, it's involuntary, meaning you don't need to direct yourself to do it consciously. Well, you're probably thinking a lot about breathing right now and it probably feels like you have to tell yourself to breathe but, under normal circumstances, your brain has it all under control. Your breathing is in cruise-control mode, so it's probably not unreasonable to forget about the fact that you're completely submerged in a fluid. Not a liquid obviously (unless you're reading this underwater, in which case, *kudos*) but a gas: air or (scientifically speaking) the *atmosphere*. The atmosphere envelops Earth and is critically important in creating the right environment for life. There's a lot of nuance to the atmosphere however; we'll unpack all of that right here in this chapter. While we're on the topic of things enveloping Earth that end in -*sphere*, we'll also dive into the *magnetosphere*. So why are the atmosphere and magnetosphere here? And how exactly do they aid our ongoing survival here on Earth? Let's take a look.

FLUID RELATIONSHIPS

When I said you're submerged in a fluid, I might have jumped ahead a little. Let's go back and clarify what a fluid is. You might use 'fluid' interchangeably with 'liquid', but that's not entirely accurate. Fluid is an umbrella term for substances that flow. So while liquid is a fluid, so too is gas. You don't really think of yourself as being immersed in, or wading through, something. Because air is invisible, it's easy to forget that it's even there, but certain things remind us we're in a big fluid. For example, why does a helium-filled balloon fly away? What has happened here? If we think about fluids, and how things of different densities behave, it becomes easier to understand. Have you ever released a tennis ball under water? If not, go do it. When you let it go, it floats to the surface almost immediately. This is because the air inside the tennis ball is lighter than the water around it. Gravity will cause lighter fluids to float above heavier fluids (or in reality, the opposite: gravity causes the heavier fluids to fill the space of the lighter object thus forcing it upwards). This same scenario is taking place with our helium balloon, except our air-filled tennis ball is now a helium-filled balloon, and our volume of water is now a volume of air. Just as our tennis ball floats upwards through the water, so too does the helium balloon through the air.

SOLIDS AND LIQUIDS AND GASES, OH MY!

You're probably quite familiar with the three fundamental states of matter already. Taking water as an example, we have:

» solid water (ice)
» liquid water (water)
» gaseous water (steam).

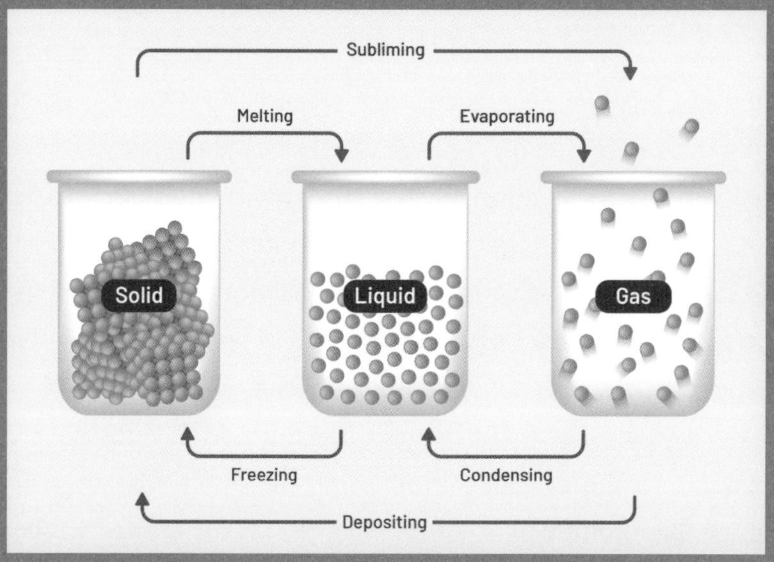

Figure 2.1: The three fundamental states of matter: solid, liquid and gas.

You may be less familiar with the fourth state of matter: plasma. Plasma is sort of like a gas, but it has been ionised. What is ionisation? If you think back to the cake example from Chapter 1, you'll recall our hydrogen atom being made by combining flour (protons) and eggs (electrons). In a plasma, the eggs have been removed, but they're not tossed away—they're free to flow around the flour separately. The nucleus of an atom (flour in our hydrogen example) and its electrons (eggs) flow freely around in this 'gas cloud'. Think of plasma as a sort of high-energy charged gas, and because the nucleus and electrons both have charges, plasma can conduct electricity and will respond to a magnet. Examples of plasma you may be familiar with are lightning, neon signs or (as the name would suggest) plasma TVs.

Fluids are important because, when they flow, they can carry stuff with them. Earlier, we discussed how important it is that water can carry things with it when it flows (for example, moving important nutrients around the body). The fluid we live in is air, and the fact that it can flow is important for us to breathe. There are a few other important aspects of an atmosphere that make it pivotal in creating an environment that can foster life, however; and it goes beyond just being able to ventilate our lungs, oxygenate our bodies and dispose of carbon dioxide.

First, let's lay down a bit of context. Look out your window. Assuming you're reading this during the day, you might be greeted by a lovely blue, sunny sky; or a dazzling display of colours at sunset; or maybe an overcast, dreary and rained-out day. Regardless, the view isn't too bad, and it's certainly not pitch-black. Now, have a look at a picture taken from the surface of the Moon, in Figure 2.2.

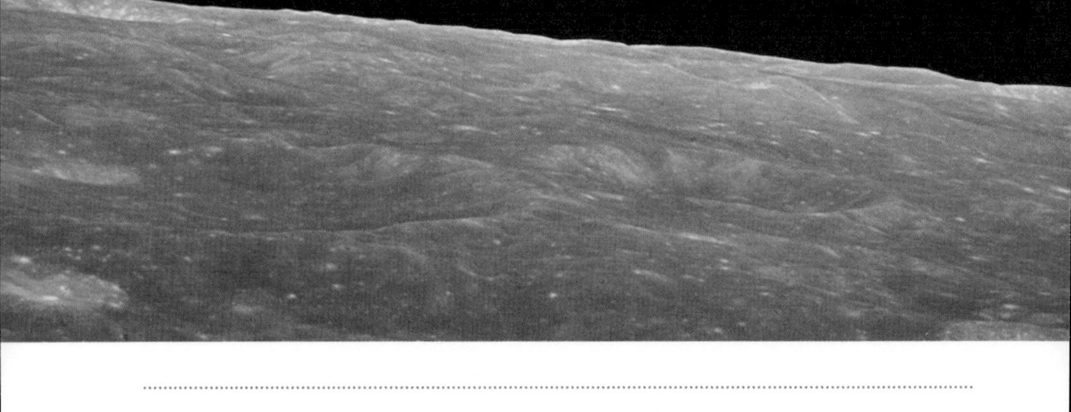

Figure 2.2: The famous 'Earthrise' photo taken during *Apollo 8*—the first human spaceflight mission to the Moon. In contrast to being on Earth, looking over the Moon's horizon yields the pitch-black void of space.

This visual is striking, right? It's humbling to see the Earth from this point of view. While we could easily sit back and bask in the beauty of this image, let's stay focused. What else stands out? I hinted at it earlier, when I mentioned we don't see pitch-black skies during daytime hours on Earth. What you can see here is that, as we look over the Moon's horizon, there's nothing but blackness. The void. The vast expanse of nothingness that we call space. The reason for this is because the Moon, unlike Earth, lacks an atmosphere.

What are the implications of the Moon not having an atmosphere? Well, we've already covered that no atmosphere means no air to breathe, but it also means we're in a real spot of bother if we're thirsty! No atmosphere means no liquid water to drink. This might seem odd, but pressure and temperature are inextricably linked. A great deal of research and experimental evidence demonstrate this relationship (Gay-Lussac's law and Avogadro's law being two examples) and it's a little tricky to describe so we'll need to go through a few steps. The first part to understand is that the change in a substance's state isn't as straightforward as we generally think. It's not quite as simple as a case of solid (ice): cold; liquid (water): lukewarm; gas (steam): hot. The *state* of a substance actually depends on both its temperature *and* pressure; it just so happens that, on Earth, the pressure is always about the same (except if you're really high up on a mountain—up there, the pressure is lower and so the boiling point is slightly different, but I'll get to that shortly). No, the relationship is quite complicated; through experimentation, we've built what is known as a *Phase Diagram* to help us understand the states of water. While we're not going to dive into the nitty-gritty of the phase diagram, essentially, it allows us to figure out what the melting and boiling points for water are for a given pressure. Assuming you made that cup of tea earlier, and you're not reading this book on a mountain, water's phase diagram will look like this:

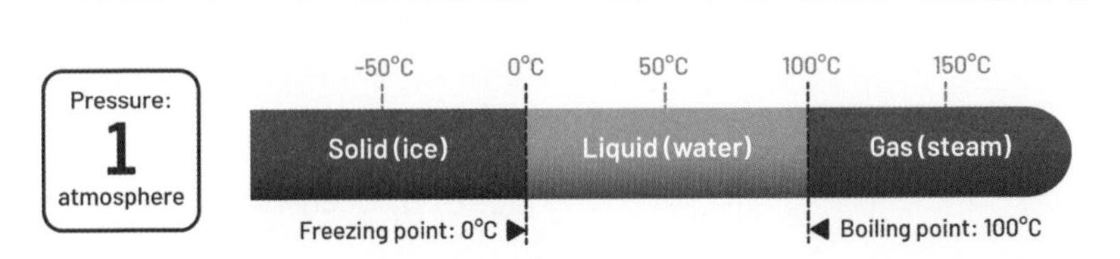

So, water freezes at 0°C, so it's a solid (ice) at lower temperatures, and water boils at 100°C, so it's a gas (steam) at higher temperatures. This is what we're familiar with. Great. Now, let's imagine that throwing down a cup of tea in the comfort of your living room isn't for you. You're more of an extreme tea-drinking kind of person. You've hiked to the top of Mount Everest, at nearly 9 km above sea level (8,849 metres to be precise!). At that height, the water phase diagram exists more like this:

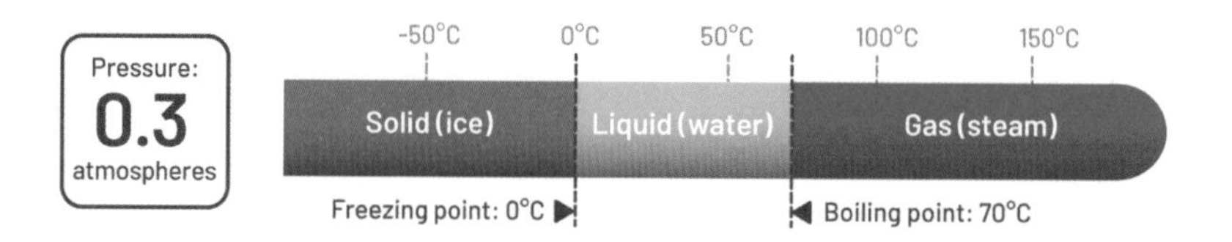

Wow, some big changes here. At the top of Mount Everest, the pressure is only 0.3 atmospheres, which means it's only about 30 per cent of what it is at sea level. That's a big drop (and why people need to use oxygen tanks so high up—the air is *really* thin). The effect this has on the boiling point of water is pretty big. Water will boil at just 70°C. So, what happens if we keep decreasing pressure? Let's create a really crazy scenario. The largest mountain in the Solar System is *Olympus Mons* on Mars; it towers over Everest, standing at a staggering height of 21 km. If Olympus Mons were on Earth, and you were getting ready to prepare your cuppa for some extreme tea drinking, water's phase diagram looks like this:

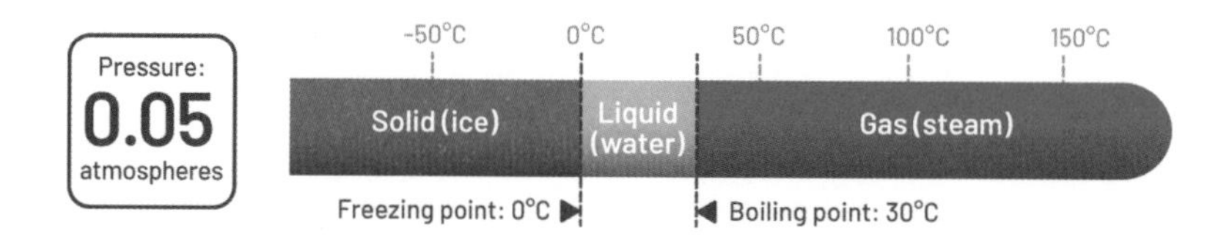

You can see what's happening here. You're not mistaken. The 'liquid' space is getting squished! As the pressure falls, so too does the boiling point, while the freezing point changes very little. At some point, the boiling point will collide into the freezing point and the area where liquid exists vanishes; here, water will only exist as a solid or gas. That sounds really weird, but you've probably encountered such a substance in real life already. *Dry ice.* Dry ice is solid carbon dioxide; at sea level, dry ice looks like this:

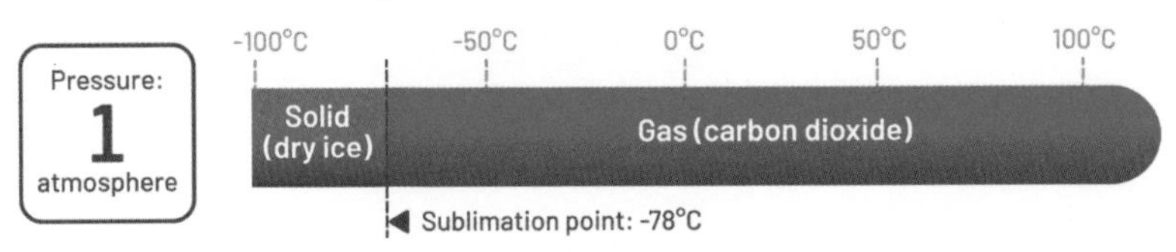

There are a few things to break down here. First, there is no liquid region. At 1 atmosphere (sea level, or your living room) carbon dioxide can only exist as a solid or a gas.

Second, there is no longer a freezing or a boiling point, just a *sublimation point.* Sublimation is the equivalent of boiling, but instead of transitioning from liquid to gas, it is transitioning from solid to gas. Water *boils* into steam. Dry ice *sublimates* into carbon dioxide gas. Finally, the sublimation point occurs at –78°C. That's right. Negative. That's ridiculously cold! Unless you're at Vostok Station in Antarctica (where the lowest-ever temperature on Earth was recorded in 1983 at an icy –89.2°C), the temperature on Earth is going to be much higher than that. As a result, dry ice immediately starts sublimating the moment it is exposed to the air, because the temperature is well above its sublimation point. Now you know how the smoke in your Halloween decorations works.

While this might seem like a bit of a tangent, it's all relevant, I promise. It's really important to understand these fundamentals in order to understand why the Moon doesn't have liquid water. We've seen how the boiling point of water decreases as pressure drops, and we've seen how, under certain conditions, substances will completely skip over the liquid phase (in our example we looked at carbon dioxide on Earth going directly from solid to gas). This is what is happening on the Moon. The lack of atmosphere on the Moon means that the surface is exposed to the vacuum of space. In essence, the vacuum of space has a pressure of zero, so the boiling point of water has well and truly collided with the freezing point of water. As such, water on the Moon will exist just like dry ice (carbon dioxide) does on Earth: it can only be a solid or a gas. On the Moon, the water phase diagram looks like this:

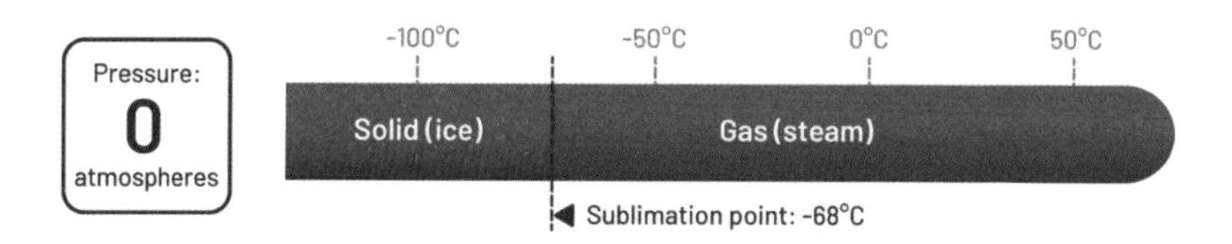

Just like dry ice on Earth, typically, any solid water (ice) on the Moon will sublimate off as gas almost immediately, and because there's a vacuum, it'll just sort of disappear, absorbed into the vast expanse of space. *That* is why the presence of an atmosphere is so important, and so critical, to life. Not only do we need an atmosphere to breathe but, oddly enough, we also need it to drink. An atmosphere acts as a sort of sleeve or shell for our planet, keeping the liquid water inside. As we discovered in the previous chapter, there's no shortage of reasons why liquid water is important, not only to our survival, but to life more generally.

ELECTROMAGNETISM

The other thing we're going to dive into in this chapter is the magnetosphere. Like the atmosphere, the magnetosphere envelops the Earth and you probably give little thought to it on any given day. It should also come as no surprise (seeing as I'm discussing it in this book) that it's important to life on Earth. What is the magnetosphere? Essentially, this is the astrophysical term for the magnetic field surrounding Earth. Before we jump into why it's important to life, let's first discuss what causes it.

We need to start from the beginning. What is a magnet? It's actually a pretty complicated answer, so we'll just stick with the most fundamental parts that are relevant to how the magnetosphere relates to life. A magnet is something that creates an invisible field (a *magnetic field*), which creates a force on select objects, notably certain metals.

Magnets have north and south poles, and the invisible field flows out of the north pole into the south pole. You've likely learned a bit about magnets during your schooling, or at least seen one in action when you've popped a picture or photo on the fridge.

Typically, you would have encountered *permanent magnets*, such as bar or horse-shoe shaped magnets (the personal favourite of Wile E. Coyote in his pursuit of the Road Runner). There's another type of magnet that's really important and provides insight into the Earth's magnetic field: *electromagnets*.

You see, electricity and magnetism are inextricably linked. In 1820, Danish physicist Hans Christian Ørsted noticed that electricity generates magnetic fields. This profound discovery eventually led to James Clerk Maxwell creating the theory of electromagnetism in 1865, combining the two separate forces of electricity and magnetism into one. It can't be overstated just how extraordinary this was.

One of the *Holy Grails of Physics* (the irony of using a religious term isn't lost on me) is the discovery of a so-called 'Theory of Everything'. This would be a theory that could neatly and succinctly describe the interconnectedness of everything; how everything works and sings together in perfect harmony. How beautiful would it be to describe the nature of the Universe in one elegant theory? Joining two previously unlinked forces (electricity and magnetism) was (and is) a huge step in achieving that goal. Electromagnetism could now be described in just four equations, called *Maxwell's Equations* (another incentive to be a scientist: do cool stuff, get your name on things). Just think about that for a moment. Everything to do with magnets or electricity can be described using just four equations. Everything. Lightning? Maxwell's equations. Fridge magnet? Maxwell's equations. Compass? Maxwell's equations. Power in your house? Maxwell's equations.

Life beyond Earth? Well, I'm writing this book about it on my computer, which is powered by electricity, and you know what describes that? Maxwell's equations! Slight tangent there, but I needed to emphasise how incredible Ørsted's discovery was.

MAGNETICS AND DYNAMOS

So where were we? Electricity and magnets are linked. And while we know about permanent magnets, what are electromagnets? This is what Hans Christian Ørsted noticed: electricity generates magnetic fields. Without going into a tonne of detail, we need to understand what electricity is in the simplest sense. Electricity is the flow of something with a charge. Typically, this is electrons (our eggs from Chapter 1) but it can also be charged particles. We can think of charged particles using our egg carton example. Remember how our cartons want to be full? That makes nature happy? Well, one way that can happen is the egg-sharing we discussed (covalent bonds), but another way is if our oxygen steals two eggs from somewhere else. In this scenario, oxygen has a nice full carton, but because it's stolen two eggs, it's got what we call a *charge*. Each egg is the equivalent of a –1 charge, so if oxygen steals two, it has a –2 charge. This is a charged particle.

When electricity flows, either with electrons or charged particles, it creates a magnetic field. And if twice as much electricity flows (say, two charged particles) in the same direction, the magnetic field is bigger and stronger. So, if we have a big, coordinated sea of charged particles, which all move around in the same direction like a synchronised swimming routine, we'd get a great big magnetic field to match. This is exactly what is happening deep inside the Earth. Earth's core is so hot that all the rocky materials are super-heated to a fluid state, and fluids can flow, remember? This fluid consists of a lot of metal elements, and has charged particles and electrons all flowing around. This flowing of charged particles and electrons creates a big, strong magnetic field that encompasses Earth. This is called the *Dynamo Theory*. Now, there's a bit more nuance to it than that, and there's a lot of complex physics taking place deep inside the Earth that makes it sustainable and stable for a really long time, but fundamentally, the flow of Earth's metal core creates a magnetic field.

WHAT'S THE GRAVITY OF THE SITUATION?

The idea behind the *Theory of Everything* is to try and combine the four fundamental forces of nature into one, unified theory. We've spoken about one of the forces, the *electromagnetic force*, already in this chapter, but what are the other three? Two of them are tricky to describe because they operate at the nuclear level: the *strong nuclear force* and *weak nuclear force*. You're probably most familiar with the fourth one: the *gravitational force*. While gravity is responsible for all things falling towards Earth, it's not just a force that makes things fall. The force of gravity attracts anything with mass to everything else with mass. Everything with mass is attracted to Earth (a really big mass!) and that's why it appears things are falling. What might surprise you is that, even though gravity impacts so many aspects of our lives (and you're potentially being gravitationally pulled into a really comfortable chair or bed while your read this), it's actually the *weakest* of the fundamental forces? And not even by a little bit. By *a lot!* Don't believe me? Pop a magnet on your fridge. Does it stay put? Yes? So that little magnet is holding on to the fridge more strongly than the *entire mass of Earth pulling gravitationally downwards*. The magnet exhibits more electromagnetic force than the whole Earth does gravitationally. Gravity is actually pretty weak in the scheme of the Universe.

THE MAGNETOSPHERE

So, now we understand how magnets work, and how electromagnetics work, we can move onto understanding why this magnetic field, or magnetosphere, is so important to life. In a rather ironic twist, the source of all energy and life on Earth also happens to be a great source of danger. The Sun bathes our little blue marble in nearly limitless energy. Life has figured out how to harness this energy and use it to thrive, reproduce and evolve. Although we're showered in usable, friendly forms of energy, the Sun is not entirely neutral in what it sends our way. We're also bombarded by the *solar wind*, a stream of charged particles and radiation expelled from the Sun. This stream of charged particles and radiation is not particularly friendly towards life on Earth. So the magnetosphere in all its protective glory shields us by deflecting this stream of particles away from us. Furthermore though, the solar wind is not only unfriendly to life on Earth, it's also destructive to our atmosphere. The solar wind would erode and deplete our atmosphere if it wasn't deflected by Earth's magnetic field and, as we've discussed earlier in this chapter, *we really like the atmosphere!* We enjoy breathing and drinking, as does the plethora of other life forms on Earth. So the magnetosphere is therefore critically important in not only helping to protect life on Earth, but it doubles down by protecting the atmosphere, too.

For all its potential dangers, the solar wind does something magical for us that it *almost* makes up for the whole 'trying to kill us' thing. While the magnetic field deflects most charged particles away from Earth, some do get through and reach us. They don't just breeze through the magnetic field, however; they still get deflected and end up just *going with the flow*. Remember how magnets have north and south poles? This applies to Earth's magnetic field as well (in fact, the magnetic poles are what your compass points towards—*magnetic north*). So what ends up happening is that some of the charged particles end up flowing along these magnetic field lines, which then enter Earth's atmosphere at the north and south poles. When they do, they smash into elements in the atmosphere and excite them. These elements get so fired up that, when they finally calm down, they have to release all that extra excitement, which they do as a dazzling display of light, shimmering and dancing in the sky. You may know this display as the 'Northern Lights' or *Aurora Borealis* in the northern hemisphere, and the 'Southern Lights' or *Aurora Australis* in the southern hemisphere. In 2013, I had the good fortune to visit Lapland in northern Finland and fulfilled my dream of seeing the aurora with my own eyes. It was an extraordinary experience, which I'll treasure. If you ever have the opportunity to view it, I encourage you to seize it! Figure 2.3 shows a photo of the Aurora Borealis as seen from Chena Lake, Alaska, as well as a photo of the Aurora Australis taken from the International Space Station.

Aside from the solar wind bombarding us with dangerous radiation and particles, we also face dangers from outside the Solar System—*really* high-energy radiation, such as *gamma rays* and *cosmic rays*. These high-energy radiation waves are incredibly dangerous to life, and the magnetic field helps to protect us from these death blasts. Hopefully, you're beginning to see that Earth's magnetic field does a hell of a lot in keeping us safe, not only protecting life from the dangers of our Sun and killer radiation from outside the Solar System, but also keeping our atmosphere safe, which we really want to keep intact.

While it's not rare for planets to have a magnetic field (the giant planets Jupiter, Saturn, Uranus and Neptune all have magnetic fields), it shouldn't be considered typical either. In fact, among the rocky planets—Mercury, Venus, Earth and Mars—only Mercury and Earth have magnetic fields, and Mercury's is only about 1% as strong as Earth's. The magnetic field (magnetosphere) is just another special property Earth possesses, in addition to many others, which all seem to click together in the right way to support life.

Figure 2.3: *Top:* A photo of the Aurora Borealis as seen from Chena Lake, Alaska. *Bottom:* The Aurora Australis photographed from the International Space Station.

CHAPTER 3

PLATE TECTONICS

Water and air. They're two of the many strings to the special bow that is Earth, and they're probably two of the most intrinsic. When it comes to our survival, breathing air and drinking water are both non-negotiables. That block of chocolate or tub of ice cream you're consuming while reading this (and if you're not, go on!) is nice, but its absence won't lead to your demise. We can (reluctantly) do without. We're not having a cheeky 'block of water' or a 'tub of air' though. We need these things to exist. They're not beneficial to our survival, they're *critical* to our survival, and so when we discussed these two aspects of what makes Earth special in the first two chapters, it made an abundance of sense. We *need* water. We *need* air. As we saw with the magnetosphere discussion in Chapter 2, however, Earth has some aspects that are integral to shaping an environment in which life can not only begin, but prosper and flourish. Beyond the magnetosphere are properties that are a little more 'invisible' to us, but are just as important in enabling life to begin.

JOURNEY TO THE CENTRE OF THE EARTH

We're going to start by talking about another of what were considered the four 'classical elements'. By classical elements, I mean those elements that ancient cultures proposed as the fundamental states of matter. At the time, these were the simplest forms of nature, from which everything else could be described: earth, fire, air and water. Of course, we have a much more complete understanding of nature and the world around us now. We define 118 elements, with even smaller objects making up these elements (such as protons and electrons), but I digress. In this chapter, we're going to look at earth, and I do mean earth with a lowercase 'e', as in the hard, rocky soil; not Earth, as in our planet. While I've stated that the concept of 'classical elements' is outdated, contemporary scientific research indicates that life does seem to place some importance on the presence of an ocean (or water, as discussed in Chapter 1), atmosphere (as discussed in Chapter 2) and landmass (as we're discussing here) co-existing to supply all the necessary ingredients for life to exist. The so-called *Habitable Trinity* appears to be super-important to worlds being habitable, and may better guide us in our search for habitable worlds beyond Earth.

First, let's have a look at the internal structure of Earth. In the previous chapter, we mentioned Earth's core. It is hot, metallic and can flow, thus allowing charged particles to flow and, in doing so, generate a magnetic field (our magnetosphere). But what else is there besides the core? Plenty, actually! We'll go through it all here, but keep referring back to Figure 3.1 for a visual reference because it can get pretty complicated.

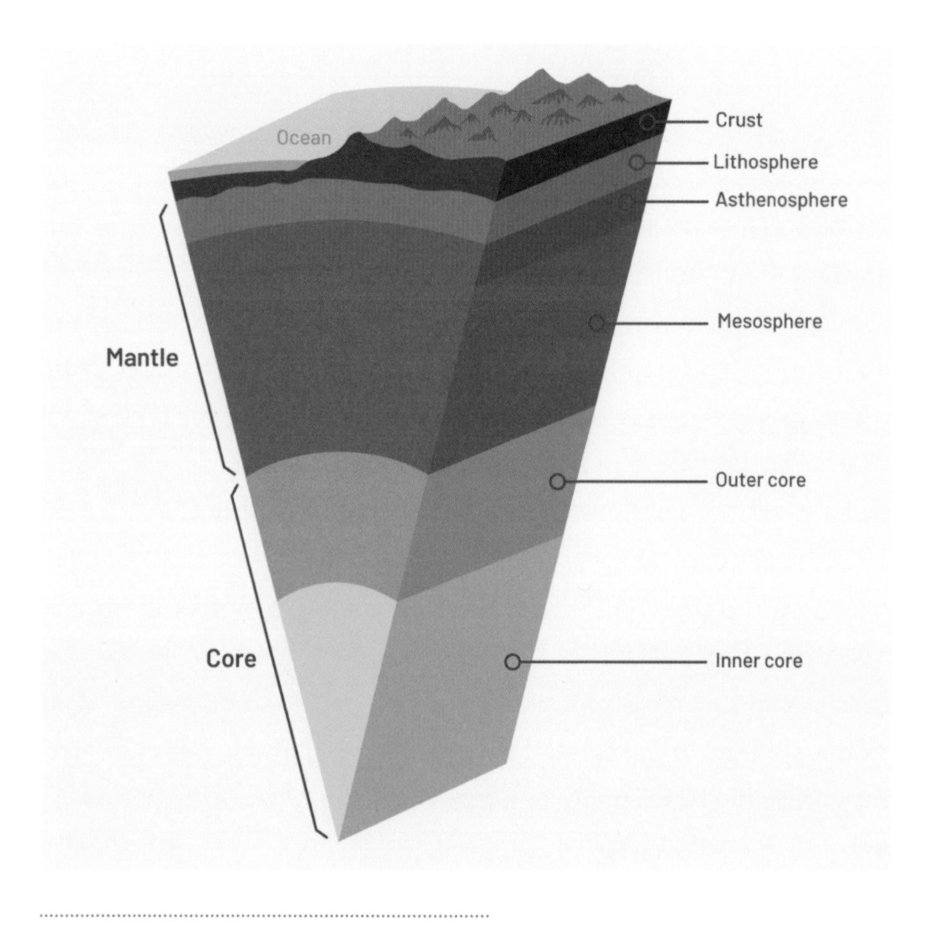

Figure 3.1: The different layers of the Earth.

Working outwards from the centre, we have the inner and outer cores. The inner core is under so much pressure from all the matter pressing down and in on itself (because gravity is pulling everything together tightly, 'down' is actually 'in'), it's not able to flow. It's being squashed too tightly. As we move further out from the inner core, however, it's still really hot but there's a little less pressure; because the outer core isn't packed quite as tightly, it can exist as a liquid and flow. Think about when you dive into a deep swimming pool. You know that feeling you get in your ears, like they're being pushed or squashed? That's extra pressure. The deeper you go, the more 'stuff' is above you (like that water in the swimming pool), which is pushing down (from gravity) and ultimately squashing things tighter. That's why the inner core is solid, but the outer core is liquid; the inner core is being squashed by more 'stuff' above it and is under more pressure, preventing it from moving or flowing like a liquid.

After the outer core is the *mantle*. The mantle is solid but, on a long enough timeline, it sort of behaves like a fluid. In case you skipped The Sounds of Celsius Fun Bubble (see Chapter 1), we discussed something called *viscosity*—in essence, a measure of the stickiness of a fluid to itself. If you pour water from a bottle, it'll flow quite freely (it has low viscosity), compared with the sluggish crawl of honey (it has high viscosity). In certain cases, some solids can actually flow, too, albeit they have such high viscosities that the flow is incredibly slow. The Earth's mantle is one such example; while solid, the material of the mantle can flow and circulate, distributing heat from the planet's core up towards the surface in a process known as *convection*. The mantle itself can be broken up roughly into two regions, the *lower mantle* (or *mesosphere*) and the *upper mantle* (or *asthenosphere*).

The asthenosphere and what lies atop comprise arguably the two most integral parts of the mechanism we refer to as *plate tectonics*. The asthenosphere is mechanically quite weak and, as we discussed above, it can move and flow. Slowly, to be sure, but it still flows. Sitting on top is Earth's outermost layer, the *lithosphere*. In contrast to the asthenosphere, the lithosphere is quite strong and rigid, which means that, while it doesn't flow like the rest of the mantle, it will slide around. The lithosphere rides on top of the asthenosphere like a surfboard—this Earthen surfing makes up plate tectonics.

SURFING THE EARTH

This gives us some context to Earth's geological structure—what Earth's interior looks like—and how the different layers interact to create the phenomenon of plate tectonics. But what is plate tectonics? Simply put, plate tectonics is a theory that describes the motion of the Earth's surface, specifically, the plates and continents on Earth. And it is the surfing plates of the lithosphere that are the basis for plate tectonics. This motion not only leads to the changing arrangement of the continents on Earth, but also the formation of many topographic features, including mountain ranges, oceanic trenches and volcanoes. These form by the oceanic plates colliding and interacting at the plate boundaries in certain ways, such as a continental collision or subduction (the process of one plate sliding underneath another plate). All these things—the motion of the plates, continental drift (the slow but measurable drift of continents across the Earth's surface), formation of topographic features, and colliding and subduction of plates—contribute to several important qualities that help to make Earth habitable and foster the right conditions for life. Let's dig a little deeper (pun intended) into some of these qualities.

Let's start with arguably the most impressive claim that plate tectonics can make: Earth's very own global thermostat! Due to Earth's internal geological activity, in the form of the core and flowing molten (read: HOT!) material, plate tectonics helps to regulate the temperature and atmosphere on Earth over really long timescales. This may seem unimportant, but the stability of Earth's atmosphere over long periods is important, and it would be impossible without plate tectonics. The process is quite complex, and is referred to as the *carbon cycle* (like we have a water cycle).

Let's see what carbon's role is in this. Carbon dioxide in our atmosphere creates a 'greenhouse effect', whereby it allows heat from the Sun into Earth's atmosphere, but traps any heat that bounces off the Earth and tries to escape into space, reflecting it back towards Earth, much like a greenhouse. The more carbon dioxide in the atmosphere, the stronger this effect is. How is plate tectonics involved in all this? The warmer the climate, the more water vapour there'll be in the atmosphere (think about Earth's tropical climates); this water vapour will dissolve carbon dioxide out of the atmosphere and it falls down as rain. Upon landing on the Earth, it can interact with rocks, dissolving some of them (erosion), and flow through rivers until it reaches the ocean. All the carbon that the water has dragged along on this merry journey can then settle onto the ocean floor, where it forms various carbon-based structures such as rocks and seashells.

Enter plate tectonics. When one plate subducts (or slides under) another, it basically drags everything on that plate with it back into Earth's interior (specifically, the upper mantle), thus removing carbon from the ocean and atmosphere.

Over a long enough time, this results in less carbon in the atmosphere and the planet cools. This means less rain, which eventually means less carbon is removed from the process described above. Because it's a cycle, however, we need to get back to the start somehow, so how does that happen? Volcanism! After hundreds of millions of years, the carbon that has been removed and subducted back into Earth's interior will flow around until, eventually due to pressure build-up or other geological activity, it bursts through the crust in a spectacular volcanic eruption, venting carbon back into the atmosphere for the cycle to continue.

In discussing the carbon cycle, we hinted at another way in which plate tectonics has created an environment for life to begin. When carbon is dissolved in water during its journey, it interacts with rock and dissolves some of these rocks. This process is critical because the elements it dissolves—copper, zinc and phosphorous—are important nutrients for organisms, specifically simple organisms such as plankton.

While this may sound more like a 'water contribution' than a 'plate tectonics contribution', it's plate tectonics that significantly fuels this process by creating mountain ranges. When plates collide and are thrust upwards, they form mountains, which are more exposed to the environment and thus allow the rain to batter the rock unfettered. This means more elements (from the mountains) dissolve in the rain water and, ultimately, flow down into Earth's oceans. What's exciting about the timing of when these plate-collision and mountain-forming processes took place—about 550 million years ago—is that they line up really well with an event called the *Cambrian Explosion*, which saw an explosion in biodiversity on the planet and a corresponding increase in organism complexity, suggesting this process is a likely contributor to the Cambrian Explosion.

THERE'S A HOLE IN OUR BUCKET

Long before the Cambrian Explosion, however, plate tectonics helped to give life another boost.

About 2.5 billion years ago, we didn't have nearly enough oxygen in the atmosphere, and the small amounts that plant life (notably, algae) produced rapidly oxidised, which is to say, it formed rust with iron-rich rocks. (Don't worry, we'll go into more detail about these changes and the Cambrian Explosion in Chapter 6. The important thing to note now is that there was little oxygen available to life.) Not having much oxygen is a bit of a problem for us oxygen-breathers, so it's fortunate that plate tectonics stepped in and helped to sort this out for us.

Here, we're going to use an analogy to describe the two effects that contributed to increased oxygen in the atmosphere. Imagine you're filling a bucket with water, but the bucket has a hole in it. In this analogy, the bucket is our atmosphere and the water is oxygen. So, the tap is a mechanism for producing oxygen (in this case, algae producing oxygen), and the hole is a mechanism for removing oxygen (in our analogy, iron-rich rocks bonding with the oxygen and rusting), thus removing it from the atmosphere. If our tap is broken and only allows a little bit of oxygen in, and our hole is big and lets a lot of oxygen out, we end up with very little oxygen in the atmosphere. Plate tectonics changes this, by varying the size of the hole in the bucket.

Over time, the concentration of iron in the crust changed due to plate tectonics. Different plates formed—for example, continental (land) or oceanic (seafloor) plates—with differing concentrations of iron. Plate tectonics led to not only the production of different plates, but also their motion, collisions and eventual subduction, where they were swallowed back into Earth's mantle. All these processes resulted in the composition of the crust above the ocean's surface (i.e. what makes up our land) to change over time, depending on how these plates move, collide and subduct under one another. Over hundreds of millions of years, the amount of iron-rich rocks exposed to the atmosphere decreased. This means that there is less iron to react with the oxygen in the atmosphere, meaning that oxygen was no longer rusting so rapidly, meaning that oxygen could exist and accumulate in the atmosphere in larger amounts. Plate tectonics, therefore, has made the hole in the bucket much smaller.

How about fixing the tap then? Even though we weren't losing oxygen as fast, we still weren't making it that quickly. Plate tectonics helped here, too. Because of the involvement of plate tectonics in the carbon cycle, eventually a load of carbon dioxide was pumped into the atmosphere (by volcanism, for example). Our carbon dioxide–breathing algae were seriously chuffed at this abundance of food. With this unfettered access to a higher concentration of carbon dioxide in the atmosphere, algae also began producing more oxygen as a by-product of photosynthesis. Simply put: more carbon dioxide food going in, meant more breathable oxygen coming out. *Plate tectonics just turned on the tap*. Now, our bucket could start filling with oxygen as we've both turned on the tap *and* decreased the hole. These steps are thought to have been pivotal in the *Great Oxygenation Event*, which saw oxygen in the atmosphere rise sharply.

Alright, so plate tectonics has been helping to ensure our atmosphere is pumped full of oxygen. Our living Earth, with its swirling flowing inner core and rigid plates surfing on top, has helped to foster an environment so life can start firing on all cylinders. Plate tectonics has been critical in oxygenating the Earth (both billions of years before and in the lead-up to the Cambrian Explosion), but how else has it contributed to creating the rich biodiversity we see today?

COMPETITIVE EDGE

As it turns out, plate tectonics is integral to arguably the most important part of evolution: *competition*. If evolution is the vehicle for getting from simple to complex life, then competition is the fuel. As life propagates from generation to generation, we see different variations on species, and because of competition for resources, only the best variations survive. Over hundreds of millions to billions of years, we end up with specialised and efficient life forms. Now, how plate tectonics fits into this is two-fold. First, the motion of the continents is responsible for creating wonderfully diverse environments, yielding a broad spectrum of habitats for life to exist. Deep underwater, plate tectonics creates hydrothermal vents on the ocean floors, where the hot mantle of the Earth spews forth, creating liquid environments beyond the reach of light, but which are still bathed in warmth and filled with the critical nutrients and elements for life to exist, which it does with aplomb. The shifting plates also create dry land masses. These allowed life to become land dwelling: evolving limbs, the ability to stand upright, and ultimately hands to craft and use tools. Different attributes are valued in these varied environments, and so different life forms succeed in each habitat.

Within these habitats, competition drives natural selection, eventually resulting in ecosystems in which each life form exists in balance. Nature finds an equilibrium. This is where the second contribution of plate tectonics kicks in. Because the continents are breaking up and re-forming, these separate, isolated environments can collide. The balance of one ecosystem can smash into another, creating bedlam. Life forms then face fresh competition for food and resources, and evolution does its thing, yielding even more diverse and varied life forms. The combination of plate tectonics creating enormously varied environments in isolation from one another, then re-forming and combining those environments, shaking up evolution again and again, is another way in which plate tectonics has shaped life on Earth and contributed to the rich tapestry of life we see today.

NATURAL CONVECTION

There's one more contribution that plate tectonics has had on Earth, which is important in creating the right environment for life. It has to do with one of the factors we discussed in Chapter 2: Earth's magnetic field. For the magnetic field to exist, there needs to be a moving electrical charge; and we find that motion in the flow of charged particles in Earth's molten core. Inside the Earth, the inner core is solid, while the outer core and mantle are fluid. Something we didn't specifically discuss is what exactly drives the motion of these fluids. Why does the liquid outer core flow at all? The reason has to do with something called *convection*—it's one of the ways that heat can be transferred from one thing to another (the other two dominant methods being *conduction* and *radiation*). Convection occurs in a fluid when there is a temperature difference between two parts of the fluid.

IT'S GETTING HOT IN HERE!

Heat gets transferred from one thing to another by several mechanisms. We've already discussed convection (the circulation of fluids of varying temperature and density), but what about the other two?

Conduction is another way that heat is transferred. It's probably the one you're most familiar with because this is how we typically feel heat with our hands (which, if you're like me, you've almost certainly burned in some way!). Conduction occurs when a hot object touches another object. At an atomic level, temperature is essentially the vibration of atoms in an object. The higher the temperature, the more the atoms are vibrating. When you touch a object, the vibrating atoms in that object start colliding with the atoms in your hand, for example. This causes the atoms in contact with the object to start vibrating, too. This is how conduction works: atoms vibrating in one object touching and vibrating the atoms in another object.

Radiation is different in how it transfers heat because it doesn't need a medium. Convection needs a fluid to circulate heat about, and conduction needs two objects to be touching. Radiation is energy that's transmitted with electromagnetic radiation; it doesn't need a medium. This might not sound like a big deal, but it's critically important because there isn't any medium in space.

If radiation did require a medium, no energy from the Sun could reach us here on Earth! Everything radiates energy, even you, but the hotter something is, the more energy it radiates. This is why you're more likely to feel the Sun kissing your face while you read this book, than the tiny amount of energy this book is radiating at you (although it's also kissing you in the face with some energy!).

Many mechanisms can drive convection, but the key one we're interested in is called *natural convection*. This arises because of how fluids change in density with temperature (if you recall, we talked about things floating and changing density in Chapter 1 with lakes freezing).

Have you ever heard the expression 'hot air rises'? This is because hot air is less dense than cooler air, so it will float above. In the same way, a bubble blown underwater rushes to the surface because the air bubble is less dense than the surrounding water. This is natural convection. If there is a hot part of a fluid, it will be less dense than the surrounding cooler parts, causing that 'bubble' or 'pocket' of warm liquid to rise. At the same time, the surrounding cooler and denser parts of the fluid will sink below the hot part. If there is a source of heat at the bottom of this fluid, then the cool sinking part will begin warming, while the rising pocket will start cooling (it's no longer near the heat source). The two will reverse; eventually, the *new* hot pocket will rise and the *new* cold pocket will sink. This results in a cycle or churning of the fluid, meaning a single heat source will cause the heat to spread slowly throughout the fluid.

We want to understand convection because it's what drives the motion of the core that creates Earth's magnetic field. The inner core is hot and heats the fluid outer core; the fluid outer core begins the process of convection, circulating and transferring heat to the upper parts of the Earth's inner structure—the upper mantle and tectonic plates. Here, tectonic plates act as a mechanism by which Earth's interior can cool. The more efficiently it cools the upper portion of Earth's interior, the bigger the difference in temperature between the cooler upper and the warmer lower pockets of Earth's fluid core. Now that the upper portion has cooled it will sink back down towards the inner, hotter core, while the next superheated portion of the Earth's interior can rise and take its place. This rotation of the hotter and cooler parts of the Earth's inner core is the very process of convection described earlier, and it is this circular motion of the fluid core that creates Earth's magnetic field. How about that? Did you expect that the plates surfing on Earth's surface were protecting you from the lethal solar wind, which the Sun is blasting your way?

Plate tectonics helps to describe many phenomena on Earth. It has played an important role in not only helping life to get started, but evolving into the enormously complex and varied life forms we see today.

CHAPTER 4

JUPITER

The Big Dog. El Grande. The Big Cheese. The BFG. The Top Banana. La Gran Enchilada. The 'King of the Planets' in the Solar System.

Jupiter. Okay, so I just made up most of those nicknames for Jupiter (although that last one has a shred of truth; it was named after the Roman god Jupiter, King of the Gods). But you get the picture: Jupiter is big! Beyond its massive size, Jupiter has played an instrumental role in helping to make Earth a special place for life. It might seem odd that another planet could have such a profound influence on the habitability of another, but it is not only an active area of research, it just so happens to be one of *my* areas of research! I have a soft spot for Jupiter because of how much time I've spent understanding and researching the degree to which Jupiter has been a benevolent protector, or malevolent destroyer, of life on Earth.

JUPITER'S BIRTH AND THE GRAND TACK

Cast your mind back to Chapter 1, when we broke down the steps of the Solar System's formation. If you recall, we only mentioned how Earth formed—dust first started to collide and stick together into pebbles, then these pebbles collided and stuck together to form rocks, and so on, until boulders, then mountain-sized objects and finally a planet formed. In essence, this is how a rocky planet forms; it's similar to how the other rocky planets—Mercury, Venus and Mars—formed. But what about Jupiter and Saturn (the gas giants), and Uranus and Neptune (the ice giants)? Planetary formation is a complex area of research, and volumes could be written about the various details of it all. The important aspect to highlight here is that, currently, the leading theory of how the Solar System formed is that Jupiter formed first. Jupiter formed similarly to the rocky planets (dust to pebble to rock to boulder to planet) but when it formed there was still a heap of gas floating about. Because of its size, this rocky object sucked up loads of the gas, accumulating more and more as a massive, thick gas blanket.

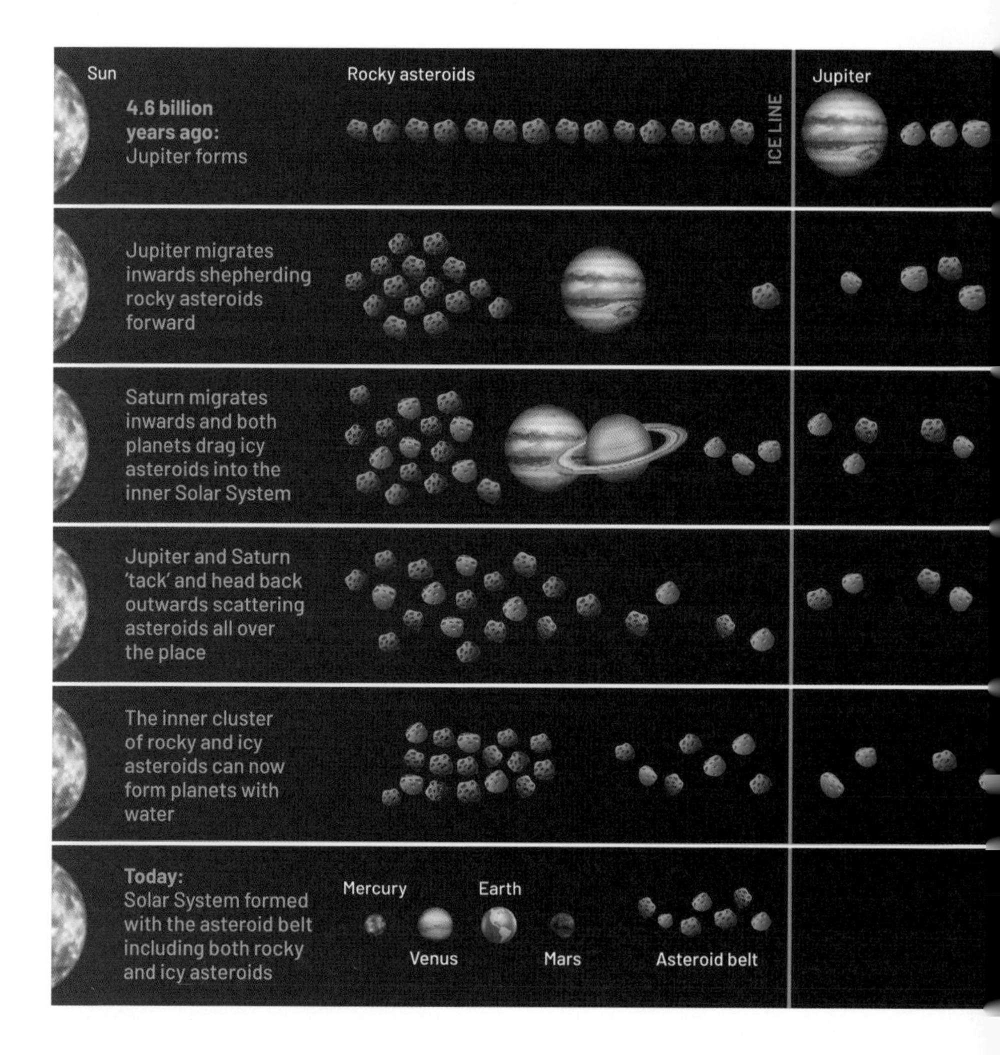

It's important to know Jupiter formed first because this gives context to the timeline in which Jupiter formed *and* did a particular manoeuvre, called the *Grand Tack*, before Earth had even formed.[2] The term 'tack' is borrowed from sailing: *tacking* is when a sailing vessel quickly reverses direction to travel against the wind. Essentially, due to the complex interactions between Jupiter and the dust-and-gas disc it was forming in, our King of Planets started to drift inwards (or *migrate*) towards the Sun.

The thing about Jupiter migrating towards the Sun is that, while it was doing so, our second gas giant, Saturn, formed and, in the same way, once its rocky core formed and it swallowed up loads of gas to

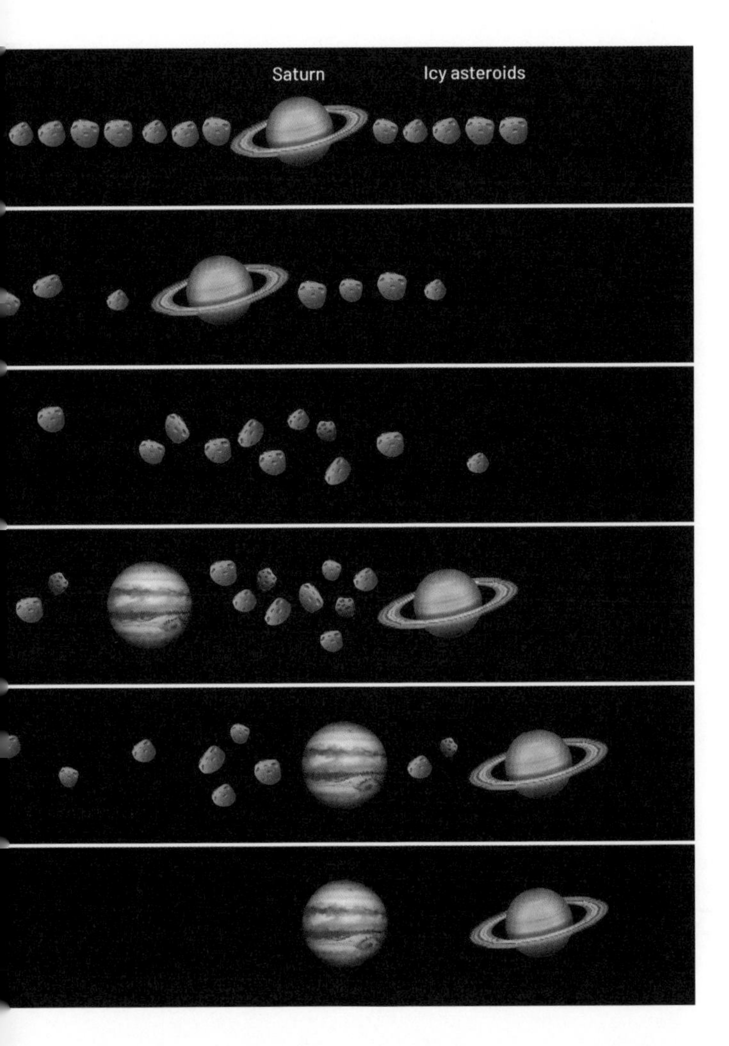

Figure 4.1: A diagram of what the Grand Tack looked like. See how more blue objects are dragged inwards to where Earth forms? That's Jupiter 'watering' our Earth. (Adapted from Walsh et. al, 2011; and Carroll, 2017.)[1]

become a massive gas giant, it too also started to migrate towards the Sun. As it turns out, Saturn was a lot faster than Jupiter; eventually, it caught up with Jupiter and performed the Grand Tack and the two planets reversed direction, moving back outwards away from the Sun until they stopped in their current orbits as we see them today.

So, why is it important that all of this happened before Earth formed? In Chapter 1, we introduced the idea of the ice line: a line around the Sun, inside of which water is a gas, and beyond which water is, well, ice. When Jupiter migrated inwards, its massive size and powerful gravitational field meant it dragged loads of icy material with it.

WHAT'S THE DEAL WITH PLUTO?

One of the questions I'm most commonly asked is: 'Why isn't Pluto a planet?' For some people, it's an emotional topic and, while it's a little misunderstood, it's also often met with dismissal: 'It was a planet when I was growing up, so it's still a planet to me.' Look, I love Pluto. I named my dog Pluto! Unfortunately (actually *very fortunately* when you give it some more thought), science is not static. We don't figure something out and then that's it locked in, forever, ad infinitum. Science grows, develops, changes and evolves. Things can be proven wrong, or modified, as new information comes to light. Remember, we used to think we were the centre of the Universe (some people still do) for a long period of time, and that the world was flat. We didn't know Neptune existed until 1846, and the theories of gravity and the atom have changed remarkably as we learn more. It's really important for science, and to our continued progress, that our view of the world changes as our wisdom grows.

So why is Pluto no longer a planet? No, it wasn't 'downgraded', it was 're-classified'; the reason is because we found that Pluto orbits around the Sun in a region of space with loads of other objects like Pluto. The options we had were to add a bunch more planets to the Solar System, or create a new classification of objects that are slightly different from planets. We went with the latter, and so, *dwarf planets* were born: planet-like objects that orbit the Sun with lots of other objects that aren't joined to another planet gravitationally. That second part is important because it rules out natural satellites, such as the Moon, from being classified as a dwarf planet.

As we discover more about how the planets formed, my guess is that we will redefine how we classify planets again. This story isn't over. Science is never finished. There's always more to be discovered, and a more complete understanding of the Universe to be found.

Gravity sort of holds onto objects, so when Jupiter moved, so did all this icy material, then when Jupiter tacked, it left the icy material swirling about in the inner Solar System. This resulted in more water in the inner Solar System than before, so when Earth finally started to form, there was plenty of water to be trapped and embedded. We've outlined a fairly complicated scenario here, so check out Figure 4.1 to help demonstrate what happened, and show how Jupiter's little dance helped to provide us with some of the vital water that life needed to start and thrive on Earth.

A FORCE OF ATTRACTION

Jupiter didn't just water the Earth before it formed. Because of its enormous gravitational field, Jupiter is a constant attractive force in the Solar System. This has two big effects on Earth. The first is that it has continued to water Earth. In the same way Jupiter dragged icy material in so

ANOTHER DWARF

Yep, a Fun Bubble inside a Fun Bubble. *We must go deeper.* Here's a fact about the object that kicked off the whole debate about re-classifying Pluto. While astronomers were finding lots of objects *like* Pluto, it wasn't until they discovered one *larger* than Pluto that they could no longer put off asking the hard questions about what makes a planet a planet. This 'larger-than-Pluto' object was eventually named Eris after the Greek goddess of discord or strife. Rather fitting, considering the planetary discord that followed!

that, when Earth formed, there was plenty of water available, Jupiter has also continued to drag in icy comets *after* the bodies of the Solar System had formed. Comets contain a significant amount of water in the form of ice. Ice is what gives them their characteristic tail: as a comet gets closer to the Sun, the thermal energy spewed out from the Sun heats up the ice so it sublimates (remember sublimation? This is where the ice turns straight into gas, not liquid). When this happens, the gas streams off the comet, carrying dust with it, thus liberating it and allowing the Sun to gently push it away from the comet.

Light from the Sun (all forms of light actually) creates a tiny bit of pressure, or force, on objects. It's incredibly small, but it's there, and is the mechanism behind what we refer to as *solar sails*. (We'll get stuck into this concept in depth in a later chapter.) The key point, however, is that this gentle push from light, acting over a long distance, on such small objects (dust particles), in the vacuum of space, all adds up to the dust being pushed off the comet and forming its long, beautiful tail.

Because the comet's tail is being pushed *away* from the Sun, it can look a bit funny and counter-intuitive as it's not directly behind the comet, like tails act on Earth (which is due to the wind dragging them back). While it has some momentum in the direction of the comet, it's being pushed in another direction, so it curves away from the Sun and back towards the path from which it came. This is in contrast to a comet's *second* tail. This second tail is usually a little fainter than its dominant tail and is tinged blue. It's made up of charged particles, or *ions*, which are strongly affected by the solar wind (that dangerous blast of charged particles from the Sun). We already know that, when charged particles move, they create a magnetic field; therefore, the solar wind must also have an accompanying magnetic field. The solar wind affects the charged particle tail much more strongly than it does the dust, which results in the blue ion tail pointing directly away from the Sun; it doesn't have the slight curve that the dust tail has. You can see this clearly in Figure 4.2—the white dust tail is brighter and curves towards the trajectory of the comet, whereas the fainter blue ion tail is straight. (While we can't see the Sun in this image, from the direction of the blue tail, we can infer the direction it must be.)

Figure 4.2: A photo of the Hale-Bopp comet taken on 14 March 1997. Two tails are clearly visible: the white dust tail (brighter and slightly curving) and the blue ion tail (fainter and pointing in a straight line).

Comets, with their dual tails and resplendent beauty, carry significant amounts of water. At times, Jupiter has dragged comets into the inner Solar System, where they've collided with Earth. This means that, if Earth didn't form with enough water (thanks, in part, to Jupiter in the first place), then Earth got a second chance to have more water delivered. This is what we see when we compare the chemical composition of the water on Earth with asteroids and comets. Certain chemical indicators in water help us understand where that water originated. Was it close to the Sun? Was it past the ice line? Was it *really far out* where comets form? Depending on the origin, the chemical indicators vary; we find that the water on Earth is a mix of several sources, including water from comets that we now know are dragged in via Jupiter. This provides us with some pretty strong evidence that Jupiter helped to make Earth nice and wet for life to begin.

Providing water to the Earth—whether before or after it formed—is not the only way Jupiter has influenced Earth. The second way is a little more contentious; it ties in with the fact that Jupiter has a strong gravitational field. Does the presence of Jupiter increase or decrease the number of objects that collide with Earth? Is Jupiter a benevolent protector of Earth, or a malevolent destroyer? The jury is still out. We've discussed that Jupiter drags objects from quite far out (where it's icy) into the inner Solar System, where these icy objects can collide with Earth. But what about larger, destructive objects that don't provide us with something beneficial such as water but, instead, cause tremendous damage to Earth and potentially the life on it? Is Jupiter to blame for the dinosaurs' deeply unfortunate relationship with an asteroid? Let's look at these two possible scenarios.

JUPITER AS THE BENEVOLENT PROTECTOR

The first scenario is that Jupiter is a protector, helping us and further ensuring our survival by 'taking the hits'. Is Jupiter our lightning rod or, in this case, an 'asteroid rod'? If you're unfamiliar with a lightning rod, it's a tall metal rod attached to the highest point of a structure that absorbs lightning strikes, protecting the structure from damage. It works because lightning seeks the path of least resistance, from the cloud where it has built up to the surface of the Earth. Because metal is a better conductor of electricity than air, lightning will naturally seek out the first metal object it can find when it descends from the sky. By extending a rod higher than a structure, lightning will seek this out. The rod can then be wired to a grounding rod safely, leaving the structure and anything (or anyone) attached to it unscathed.

So what is this 'asteroid rod' role that Jupiter may fulfil? It all comes back to Jupiter's massive size and accompanying gravitational field. If a large object went rogue and somehow ended up on a collision course with Earth, in the 'protective Jupiter' scenario, the object might be intercepted by Jupiter or, more specifically, Jupiter's gravitational field. The object could be dragged into Jupiter's orbit; it could get caught, spiral around Jupiter and eventually collide into it; or it could even slingshot around, like when you putt a golf ball in mini golf and the ball skirts around the edge of the hole before shooting off in a different direction.

This same scenario could happen with a large object skirting around the edge of Jupiter's gravitational field: it's not close enough to be caught in orbit, but it's not far enough away to pass by, so it gets 'deflected' from its original path. In either of these cases, however, the object is no longer on a collision course with Earth. With its massive size and far-reaching gravitational field, Jupiter can capture large objects, which then impact on itself, or deflect the objects away from Earth. In this scenario, perhaps Jupiter gave the dinosaurs a longer innings on Earth by stopping earlier collisions but, eventually, one 'got through to the keeper'.

JUPITER AS THE MALEVOLENT DESTROYER

In the 'destructive Jupiter' scenario, instead of Jupiter dragging in only smaller, icy bodies from the outer Solar System, it can also drag in larger destructive bodies. More importantly, Jupiter can also dislodge objects already inside the inner Solar System, rendering them unstable and setting them on chaotic orbits where they could end up anywhere. The worst being right in our faces!

It might seem a bit weird to think that Jupiter can not only pull objects from the outer Solar System inwards, but can also sort of 'push' objects from the inner Solar System even further inwards until they collide with Earth. It comes down to what's called *orbital dynamics*—the dynamics or motion of objects in orbit due to gravity. If gravity was fixed and constant, such as it would be if no other planets existed in the Solar System, then objects would just orbit the Sun like clockwork. Beautiful. Regular. Perfect.

The Solar System isn't perfect, however, and it isn't constant because of the Big Cheese, Jupiter (all the planets, actually, but Jupiter has the biggest effect). The motion of Jupiter can change the gravity for planets in the inner Solar System, because the Sun is tugging one way with all its gravity and Jupiter is tugging another way with all its gravity. This *gravitational tug-o-war* can cause an object to fly off its nice, stable orbit into a different orbit. It could be one that ultimately crosses over the orbit of Earth and then: *BOOM!*

So which is it? Is Jupiter a goodie or a baddie? A hero or a villain? Could Jupiter have behaved in both ways at different times? Well, we simply don't know. The Solar System has already formed, so we have to try to figure it out by:

1. hypothesising what could have happened
2. figuring out what we would be able to observe if that hypothesis was true
3. trying to observe whatever that particular observation is.

At the point of writing this book, we're still piecing it all together ... but you there, reader from the far off distant future, you already know the answer, don't you?

DINNER PLATES AND SPINNING TOPS

There's one final way that Jupiter can influence Earth and its ability to harbour life: its contribution to *Milankovitch cycles*. This is an umbrella term for how subtle changes in the motion of Earth result in large changes in Earth's climate over the course of thousands of years. To understand what that means, first, we need to discuss a few key things regarding Earth and its orbit around the Sun, so let's dive right in.

Earth travels in a big circle around the Sun, right? *Almost*. For a lot of things, we can indeed approximate Earth's orbit to a perfect circle but, in reality, it isn't quite perfect. There are several parameters to a planet's orbit, but I'm only going to discuss two here. The first is called the *semi-major axis*. In a perfectly circular orbit, the semi-major axis is the circle's radius (distance from the centre of the circle to its edge). The reason we call it the semi-major axis and not the radius, however, is because orbits aren't perfectly circular. They are actually oval in shape, and with ovals, you have two distances.

Picture an Australian Football League (AFL) oval. The longer distance corresponds with the distance between the goalposts at each end of the field, while the shorter distance corresponds to the distance *across* the field, from the boundary on the wings. In this scenario, the distance from the centre of the oval to one set of goalposts is the semi-major axis, and the distance from the centre of the oval to one of the wings is the *semi-minor axis*. So, the semi-major axis tells us how big an orbit is; how far away from the Sun the planet is at its furthest point.

The second parameter is called the *eccentricity*—this refers to how much of an oval the orbit is. Figure 4.3 shows objects with different eccentricities. An eccentricity of 0 means a perfect circle. An eccentricity of 0.3 might be like a slightly squashed ball, while an eccentricity of 0.9 would be like an AFL football (the ball itself, not the oval).

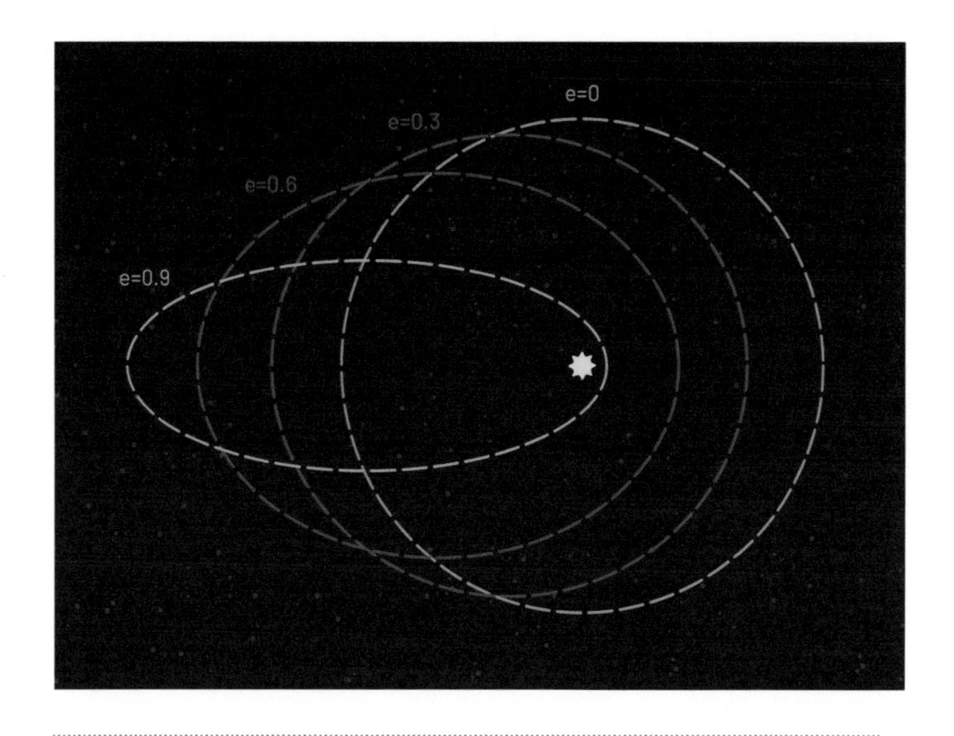

Figure 4.3: A planet's orbit is different for different eccentricities. Notice that, as an orbit's eccentricity increases, there is a 'near' side of the orbit (where the planet is near to the star) and a 'far' side (where the planet is far away).

In addition to these two values that describe Earth's orbit, I also want to highlight that Earth is tilted on its axis. What this means is that, while Earth spins on its axis (each rotation is one day), it is also orbiting the Sun (each completed lap is one year). Earth's rotation isn't perfectly lined up with its orbit. If the orbit is a dinner plate and our Earth is a spinning top, the top isn't spinning perpendicularly with the surface of the plate, it's slightly tilted.

I wanted to introduce these ideas of eccentricity and Earth's tilted rotation because they're integral to understanding Milankovitch cycles, and the link between them is called *precession*. Precession refers to the change in orientation of a rotating body—this is a lot of words to take in, so let's go back to our spinning-top analogy.

Have you ever spun a spinning top and, once it's slowed a little, it keeps spinning but it also wobbles about? That's precession! The top is still spinning around but, as it starts wobbling about, it's changing its orientation. That happens with Earth, too—it keeps spinning but it also wobbles about. The 'top' of the 'Earth spinning top' moves around. The difference here is that, while you can see a spinning top wobble and precess when you spin it, Earth does its wobbling over *long* periods of time. We're talking 25,000 years here, so you'd never notice any change in your lifetime.

This phenomenon occurs not just with Earth's rotation, but also its orbit. We've highlighted that Earth's orbit is slightly oval shaped because we don't live in a perfect, circular, zero-eccentricity world. Looking back to Figure 4.3, you can see the star and the oval-shaped orbits, and that the orbits are oriented in such a way that one side of the orbit is nearer to the star (the right-hand side), and the other side is further from the star (the left-hand side). This gives the orbit an orientation, meaning it can also precess. The orbit will gradually spin around the Sun. This precession takes even longer, about 112,000 years. You definitely won't notice this in your lifetime.

Jupiter also impacts the eccentricity of Earth's orbit. How much of an oval is Earth's orbit? Because Jupiter stretches and squashes the orbit, the difference between the near and far sides of the orbit will also change as Earth's orbit oscillates between being nearer to a circle or nearer to a football. This takes a long time, too, about 100,000 years.

JUPITER'S IMPACT ON EARTH'S CLIMATE

So where am I going with all this stuff about precession, rotating orbits and orbit shape? These phenomena all contribute significantly to Earth's global climate and, in some cases, they are responsible for glacial periods or *ice ages*. More specifically, Jupiter causes changes in the eccentricity and precession of Earth's orbit; the subtle pull of Jupiter's great big gravitational field as it moves around the Solar System causes the shape of Earth's orbit to change and precess. Would you have ever thought that Jupiter, a planet more than 600 million kilometres away, could be contributing in some way to Earth's climate? I guess the bigger question should be: Are the Milankovitch cycles that Jupiter causes good or bad for life on Earth?

That's the thing about Jupiter—as the biggest, baddest planet in the Solar System, its ability to influence the other planets is unmatched. While we don't know whether Jupiter has an ultimately beneficial or detrimental effect on helping to nurture the right environment for life, there is one compelling reason to believe it's the former: we're here. We have exactly one example of life beginning and thriving on a planet, one piece of evidence, one data point, which is that we're here, and our Earth has existed the whole time Jupiter has been doing its thing. So, if Jupiter has significantly influenced the environment on Earth, and life has been born and allowed to thrive, then it seems to follow that Jupiter has helped make Earth a special place.

CHAPTER 5

THE MOON

♫ *When the Moon hits your eye,*
Like a giant, regolith-coated, oxygen-silicon-iron dominant natural satellite of Earth,
That's amore. ♫

Okay, so it's not nearly as catchy nor easy on the ears as Dean Martin's rendition, but we've squeezed a little more detail about the Moon into the lyrics than just the circularity it shares with a pizza.

Most nights, the Moon puts on dazzling displays for us to enjoy. Sometimes, it's the beautifully illuminating full moon, sometimes, it's the cheeky Cheshire cat smile of a crescent moon. These phases of the Moon, and everything in-between, offer us a pretty show to enjoy every night. As children, we may think of the Moon as being the Sun's counterpart: they're the King and Queen of the sky. Just as the Sun rules the daytime sky, the Moon rules the night. However, the two are strikingly different in almost every way. They're not counterparts at all. They formed differently, evolved differently, have different properties and contribute to the environment on Earth *very* differently—it is this last point that we will discuss here. More specifically, how has the Moon helped (and still does) to support life on Earth?

HOW DID THE MOON ACTUALLY FORM?

Let's start from the beginning. How is the Moon different from the Sun, and how did it form? We touched on the Sun's formation in Chapter 1 when we looked at how the Solar System formed: the Sun formed as dust and gas were collapsed by gravity, which led to a greater and greater mass of material collecting in one spot. Eventually, there was so much mass in such a small volume that nuclear fusion began; this is the process in which two or more elements are slammed together to form a heavier element, releasing *loads* of energy in the process.

That's the quick recap on the Sun, so what about the Moon?

Well, the Moon is completely different from the Sun, other than appearing to be a similar size in the sky. The Moon isn't a star or relative of a star. It is more similar to a planet, both in its chemical composition and how it formed. Just like any discussions we have about how the planets of the Solar System formed, there are several competing hypotheses about how the Moon formed. And understandably so, right? It's not easy trying to figure out what happened over the last 4.6 billion years, purely from what we see today.

Some hypotheses are much stronger contenders than others, however, because they explain many observations we see today. We'll go through the most widely accepted one here for how the Moon formed, named (appropriately enough, as you'll soon find out!) the *giant-impact hypothesis*.

THE COSMIC BALLET GOES ON

Okay, so the Moon and Sun formed differently, they're made up of different stuff, and they differ enormously in size. They look the same size in the sky though, so what gives? It just so happens that the Sun is about 400 times larger than the Moon (401 times larger, to be more precise), but it is about 400 times further away from us (387 times, to be more precise). Combined, these two effects make it appear that the Sun and Moon are the same size in the sky. To test this out, here's a little experiment. If we approximate the length of an AFL field to be 100 metres (here's the scientist in me: approximate everything as neat multiples of 10—it makes the maths easier), and you're sitting behind the goals (not too far back). If you then hold a $2 coin about 30 cm in front of your face, then it should be about as wide as the goal posts. Try it next time you're at a game!

The most wonderful thing about this little coincidence? Because the Moon is about the same size as the Sun in the sky, it can block out the Sun if it crosses in front of it. This is, of course, a solar eclipse. It's so spectacular because the Moon and Sun are so similar in size. There are plenty of nifty ways to observe one safely, so make sure you're ready the next time one is viewable wherever you are.

CLASH OF THE PROTOPLANETS

Let's set the scene: nearly 4.6 billion years ago, the Sun formed and started spewing light and energy out towards the swirling mass of gas and dust that orbits it. The gravity of the Sun pulling inwards—as well as some complex physics that caused the gas and dust to mix, clump up, drag and behave in a variety of ways—caused larger and larger bodies to form. The big bad gas giants (Jupiter and Saturn) and ice giants (Uranus and Neptune) formed first, with the rocky planets (Mercury, Venus, Earth and Mars) gearing up for their turn. At some stage on the journey from dust particles to planet, a planet-like body reached a size that wasn't *quite* the size of a planet, but was on its way. We refer to this as a *planetary embryo* or *protoplanet*. A bit like a *pre*-planet.

When the rocky planets were getting ready to form, a bunch of these protoplanets were orbiting in the inner Solar System and most would eventually merge and combine to form a handful of larger bodies: the rocky planets. It was about 150 million years after the Sun formed that the whole 'Moon formation' thing went down. A pretty big protoplanet had already merged with a bunch of other bodies orbiting the Sun: *proto-Earth*. Unfortunately for proto-Earth, there was another pretty big body in the same orbit: *Theia* (I'll get to the name shortly). Theia was about the size of what Mars is today, so not as big as Earth, sure, but it was *big*.

Due to all the chaos in the early Solar System, and the competing gravities of the Sun and Jupiter, Theia was nudged out of a more stable orbit and slammed directly into proto-Earth. As you'd expect, this created an enormous impact. Loads of material from the two bodies were vigorously mixed together, material was dislodged from both proto-Earth and Theia, and all sorts of mayhem took place. Fortunately, there were still tens of millions of years before Earth formed, which was plenty of time for that chaos to calm down and subside.

With enough time, proto-Earth formed into the Earth—the planet returned to a smoother, spherical shape and, most importantly, all that material dislodged from the impact of Theia and proto-Earth that wasn't violently expelled from the grasps of the Earth's gravitational field remained in orbit and coalesced together to form one large body in orbit around Earth: the Moon. That's how we think it all went down. Now, getting back to the name, Theia. Theia, in Greek mythology, was the mother of Selene, the goddess of the Moon, just as Theia was the mother of our Moon. It's a bit cute, isn't it?

So that's how the Moon came to be. Because of that formation process, there is something quite special about the Moon: it's big. Okay, I should be more specific: it's big *relative to the planet it orbits*. To clarify, the Moon isn't the biggest natural satellite (or moon—lowercase 'm' when talking about natural satellites, uppercase 'M' when talking about *our* natural satellite) in the Solar System. That title is held by Ganymede, one of the moons of Jupiter, which is larger than the planet Mercury (not by much—if Ganymede was the size of a cricket ball, Mercury would be a tennis ball). The Moon isn't even in the top three—it's the fifth-largest moon in the Solar System.

THE DARK SIDE OF THE MOON

For many of us, 'The dark side of the Moon' immediately conjures up images of light refracting through a triangular prism into a rainbow, à la Pink Floyd's eighth studio album. It's also sometimes used to refer to the far side of the Moon. Here, *dark* is being used in the same way it's used when referring to *dark matter*, meaning 'unknown or unseen'. In reality, there is no dark side of the Moon that is completely devoid of light. Just like Earth, the Moon rotates so the entire surface of the Moon is illuminated at different times of the lunar day.

The far side of the Moon refers to the side that doesn't face us; weirdly, the side that doesn't face us is always the same. This is a phenomenon called *tidal locking*, which means that the Moon orbits and rotates Earth at the same speed. So we only ever see one side of the Moon. Go out and look at it now. Or look at a photo of the Moon taken from Earth. You'll always see the same face, craters and features. At times, it will be completely illuminated (a full moon); other times, it will be partially illuminated (a crescent or gibbous moon); and sometimes it won't be illuminated at all (a new moon). We'll explain tidal locking in more detail in Part 2.

WHAT'S THE (DARK) MATTER?

Because we glossed over *dark matter* earlier, I just want to clarify what it is here. In this case, *dark* means 'unknown or unseen', and *matter* is basically stuff; anything that has mass. There's a little more nuance to it, but that's the basics. So, dark matter is invisible stuff. Okay then, how do we know it's there, and why do we care? One of the biggest clues comes from observing galaxies spinning. Based on how much mass we can see in spiral galaxies, and how fast they're spinning, the arms of galaxies spin way faster than they should. Think about the Solar System. The planets further out orbit more slowly than the inner planets (Jupiter moves a little under half the speed of Earth, for example). If a galaxy behaved similarly (that is, with most of the mass in the centre of the galaxy where the bright, tightly packed group of stars is), then the arms should also rotate more slowly, like the outer planets of our Solar System.

But they don't. They orbit at about the same speed as the inner stars. The way to solve this problem is for a bunch more mass to exist that we can't see: dark matter. Why should we care? Because there's about six times as much dark matter as there is matter, and *we don't even know what it is!*

Relative to Earth, however, the Moon is big. It's about 27 per cent the size of Earth. In contrast, the Solar System's four biggest moons are Ganymede (4 per cent the size of Jupiter), Titan (4 per cent the size of Saturn), and Callisto and Io (both 3 per cent the size of Jupiter). The Moon is so big relative to Earth that some scientists think about the Earth–Moon system as being a double planet. While the definition of a double planet isn't well defined, as we discover more systems beyond our own, we may get some clarity about whether the Earth–Moon system is more of a planet–satellite or planet–planet system.

LUNAR INFLUENCES

This is all well and good, but how does this affect us here on Earth? How does the Moon help in supporting life? The Moon has influenced and continues to influence the environment of Earth in several significant ways (its past influence was even stronger because, when it formed, the Moon was closer to Earth than it is now—yes, it's slowly drifting away from us, but fear not, it only drifts about 4 cm each year).

The first influence is the Moon's stabilising effect on Earth. As we've discussed, the Moon is pretty big compared to Earth. It also happens to be pretty close. This results in the Moon's gravitational pull tugging on Earth. Not so much that it causes absolute mayhem, but enough to have an effect on it. And the effect is really useful: it helps to stabilise Earth. And that's pretty important with that big brute Jupiter orbiting a little further out. The Moon dampens some of the intense gravitational effects of Jupiter.

Just as your car's suspension helps to cushion its abrupt motion as you travel over bumps in the road, the Moon helps to prevent Jupiter and Saturn gravitationally disrupting Earth. This means that the rotation of Earth's orbit is really stable, specifically the tilt of the Earth. In the previous chapter, we discussed the idea of precession, using the example of a spinning top. Let's use the spinning top again to describe Earth's axial tilt. The tilt is the angle from a perfectly vertical line to the position of the handle of the spinning top. Figure 5.1 shows what this looks like on both a spinning top and Earth. On Earth, the tilt is usually between about 22 degrees and 24 degrees, and is the cause of the seasons.

Now, without the intervention of the Moon, the aforementioned gravitational influence of Jupiter on the Earth would cause this to vary significantly more, which wouldn't be great because we'd have a really chaotic climate. Earth's climate would be more extreme, and could prevent life from evolving to more complex life forms. A stable climate over millions of years gave evolution plenty of time to work its magic. A violently variable tilt on its axis would prevent such stability. Fortunately for us, the Moon cushions the blow and prevents the gravitational tug-of-war from significantly changing the tilt, thus leaving a lovely, stable climate that ultimately (over a long time) resulted in us.

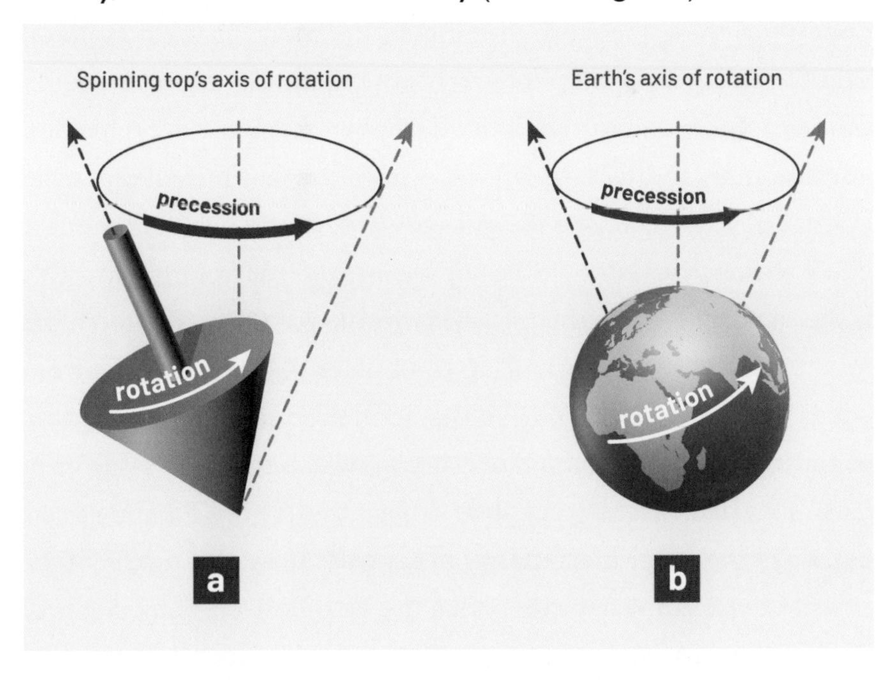

Figure 5.1: The axial tilt of (left) a spinning top and (right) Earth.

THE MOTION OF THE OCEAN

The second influence the Moon has on helping life to exist on Earth is arguably the most obvious: the tides. I'm sure you've heard that the Moon affects the tide, but how does that work exactly? Well, the fundamental idea has to do with the fact that gravity decreases as the distance from an object increases. The gravity of the Moon exerts a stronger pull on the side of Earth that's nearer to the Moon than on the opposite side. Because Earth is solid and rocky, it will remain the same shape even when subject to these differences in gravity, but what about the water sitting on Earth's surface? That gets pulled harder on the Moon side than on the non-Moon side—it's *tugged* in the direction of the Moon.

On the non-Moon side, a different phenomenon occurs. The gravitational pull of an object is stronger the closer something is to it. This means that just as the gravity of the Moon pulls the water on the Moon side of the Earth more than it pulls the Earth itself, so too does it pull the Earth more than it pulls the water on the non-Moon side of the Earth. This has the effect of creating a second bulge on the non-Moon side of the Earth. This phenomenon can be hard to visualise so try this: fill a glass of water to halfway and place it on a table in front of you. Now, slide the glass towards you in one swift motion. You'll notice that water sloshes up against the back of the glass. This is the same kind of idea. In this scenario, the glass is Earth, you're the Moon and the water sloshing up the side of the glass furthest from you is the ocean bulging out on the non-Moon side. This means that water on both the Moon and non-Moon sides is stretched away from Earth's surface, resulting in bulges or *high tides*. Figure 5.2 shows this bulging of the water more clearly. In contrast, the water on the other sides of the Earth won't experience the Moon's strengthened gravity or the centrifugal force and so will sit at a lower sea level, resulting in *low tides*. Ultimately, we have periods of low and high tides twice a day, depending on where the water is located on Earth relative to the Moon. The forces that create these tidal effects are called *tidal forces.*

The oscillating periods of low and high tides create the motion and flow of water in the ocean. They cause some fairly important things to take place such as tidal currents, which are vital for the distribution of heat from the equator (where it's warmer) to the poles (where it's cooler), as well as the redistribution of nutrients and toxins around the planet. The tides have an impact on the ocean's temperature, the availability of resources and (most importantly) portions of the land experiencing alternating periods of being exposed and submerged. This may have been particularly important for some of the earliest life forms, especially for that first profound jump from sea to land!

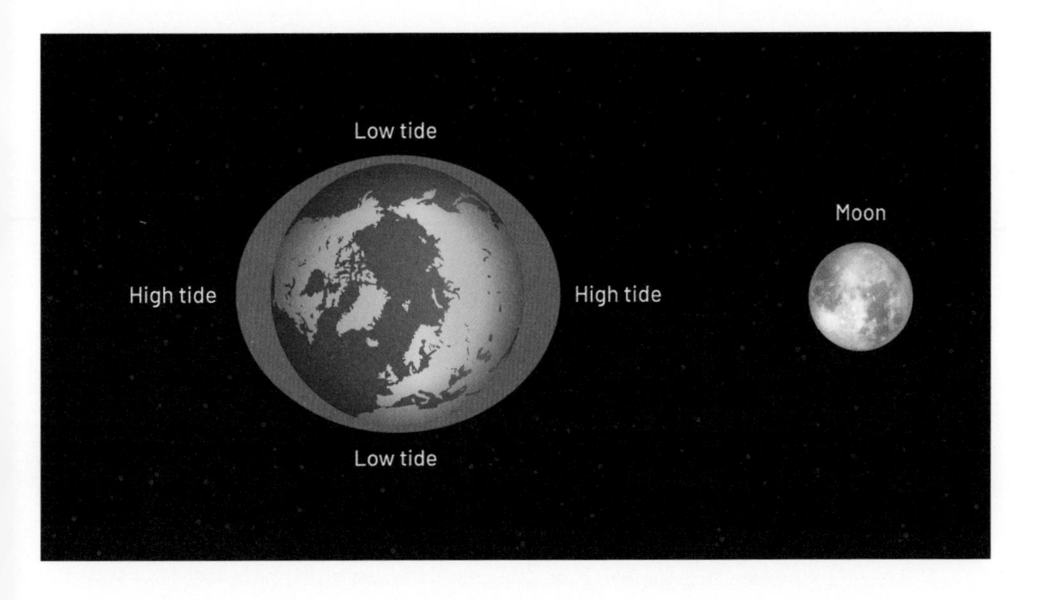

Figure 5.2: This diagram illustrates how the Moon creates two high tides (on the sides nearest to and furthest from the Moon), and two low tides (on the other sides of the Earth).

RULER OF THE NIGHT

The Moon has not only been instrumental in enabling a plethora of biodiversity to evolve in the past, but its significance to life on Earth is still evident today. Remember how we talked about the Moon being the ruler of the night? While it is strikingly different from the Sun, it does offer one of the same benefits, albeit in a diminished capacity: light. The Moon doesn't emit light itself, but reflects light from the Sun back towards us. This great celestial mirror is a thing of beauty, to be sure, but it also means that night-time on Earth is alive and bustling with activity.

Life forms have developed a myriad of nocturnal adaptations, allowing for life to thrive even during Earth's witching hours. Some life forms have evolved such that the Moon is critical to their continued existence; these organisms have grown to rely on the Moon for navigation and procreation. Navigation is important if you need to make a quick escape, such as for the African dung beetle. After stumbling across some animal excrement and committing a faecal smash-and-grab, the African dung beetle needs to hightail it out of there, and the last thing it needs is to get disoriented and have another beetle commit a manure mugging. This is where the light of the Moon lends a helping hand: the African dung beetle utilises this light to travel in straight lines, thus ensuring the security of the beetle's acquired number-two is priority number-one.

For some animals, procreation is intrinsically linked with the Moon. The most profound and awe-inspiring example of this is the coral polyps of the Great Barrier Reef, releasing their eggs and sperm. The Moon triggers this massive spawning event, which is so big it can be seen from space! And it's not just sea life, either. Barau's petrels—seabirds that breed largely on Réunion island, in the Indian Ocean—rely on the full moon to synchronise their arrivals so they can breed. While it's nice to think that the Moon's beauty and romance enchant these birds, it's more likely that the animals can interpret the Moon's subtle cues to ensure their continued existence.

While the Moon is beautiful, it's also pretty important. Since Theia collided with Earth and formed the Moon, the two have been tightly coupled, so it's no surprise that the Moon has impacted life the way it has. Its presence has helped to stabilise Earth, created environments that foster life and enabled adaptations to evolve—clearly, the Moon has been (and continues to be) far more critical to life on Earth than we first thought. It may also be crucial for our continued existence in the future in a different way.

In 1969, we (humanity) succeeded in putting a human on the Moon, just 66 years after the first powered aircraft flight on Earth (that's pretty mind-blowing to think about, but I digress). The Moon is the closest celestial object to us, and it may become a station for us to continue to explore, and establish ourselves within, the Solar System. The Moon's gravity is much lower than the Earth's, so it's much easier to leave it; we aren't battling against as much gravity. The Moon's proximity to us makes it a logical first step to establish a self-sustaining, semi-permanent colony for future missions to prepare, consolidate resources and launch from using much less energy.

The Moon is also abundant in mineral resources; this may give us an easier pathway for future space engineering rather than hauling materials into space all the way from Earth. This might sound all very science fiction-y, but remember, the International Space Station (ISS) has had continuous human habitation for more than 20 years. Science fiction and reality have collided already! Regardless, the Moon has been important to life on Earth, particularly to us. It's almost certainly going to be integral to our continued exploration of the Solar System.

CHAPTER 6

LIFE

It may seem obvious, but that doesn't make it any less true: Earth is the only place we know where life exists. Life doesn't only exist here, but it flourishes and prospers. In the previous chapters, we discussed some of the things that make the environment of Earth ideal for fostering life, but that's only half the story.

How did life start? How did it change? How did we end up with the wildly rich and diverse life forms that exist across the planet? These questions are part of some of the biggest mysteries in science so, unfortunately, you won't learn how life began in this book. As a species, we are still figuring that one out. Instead, what we will go into detail about here is some of the other steps. Once life began in its most simple form, what happened next? How did we go from the simplest life forms imaginable, to us learning about those life forms in this book? Let's find out!

WHAT EVEN IS LIFE?

First things first: What is life? When you think about it, it's not nearly as easy to define as you might think. Intuitively, we know when something is alive or not (whether because it never was, or has ceased to be, alive) but, in science, we need something more formal than that. Fortunately, an entire branch of science is focused entirely on the study of life and living things: biology. Despite this, it's still really hard to define life formally. Hundreds of definitions have been formulated, some elegantly simple, others complicated. Nevertheless, some properties pop up more frequently than others. Typically, life will exhibit all or most of the following:

1. **Reproduction** This one's simple enough. Reproduction is the ability for life to procreate, replicate or pass on its genes in some capacity.

2. **Homeostasis** This is the ability for life to maintain a constant, living state. Consider your own body—its temperature should typically sit at about 37°C. This is whether you're standing in the snow, swimming in the ocean or walking in the sunshine on a scorching hot day. Your body will seek to maintain a constant temperature, which helps it to carry on living.

3. **Growth** Life has a cycle. Organisms will come into being, develop, mature, pass on their genes and then cease to be.

4. **Energy/metabolism** Life needs energy. We humans achieve this through our metabolism: the process by which we convert the food and drink we consume into energy our bodies (or more specifically, the cells of our bodies) can use. Other life forms, such as plants, achieve it through photosynthesis: a process that converts sunlight into energy their cells can use.

5. **Response to stimuli** Life does stuff. This stuff can be quite complex, such as conversing with one another, laughing out loud at someone's hilarious joke, or the fluttering of your heart when a crush smiles at you, but it can also be much simpler. It can be sunflowers gently turning to face the Sun, or an African dung beetle (as mentioned in Chapter 5), navigating a straight path using moonlight.

6 **Evolutionary adaptation** Just as life responds to stuff (or stimuli) immediately, it also responds to the environment more gradually over time. Life mutates and evolves, leading to competition among different forms of life, where those mutations better suited to the changing environments win and those less suited lose (we'll go into more detail about the nuance of evolution later).

7 **Organisation/order** Just as a house is made up of many bricks, life is made up of many molecules, specifically arranged into one or more cells, that has a structure, or order, to it.

This is by no means an exhaustive list—nor does life have to tick every box—but it will suffice for the purposes of our discussion.

Okay, now we know what life is, at least for our purposes, let's start from the beginning.

HOW DID LIFE BEGIN?

How did things begin? And how did they eventually make it to the situation in which we find ourselves today: with the seemingly boundless biodiversity—from the simplest single-cell organisms to the extraordinarily complex—all coexisting in equilibrium? These aren't easy questions to answer, and we're certainly not going to be able to explore them in nearly enough detail here; so instead, let's look at a slightly abridged version of events, focusing on the most relevant parts—just the Cliffs Notes for the origin of life. With that in mind, let's dive in.

KEEP YOUR IRIS ON THE VIRUS

When we think of a virus, we generally think of something that isn't particularly good for us. The common cold, influenza, human immunodeficiency virus (HIV) and rabies are just some of the viruses that affect humans. As you can see from the small sample I've listed, they can vary from minor inconveniences, to something that should be managed with medication, to death. One of the reasons viruses are so problematic to treat and combat is because, well, they're not *really* alive. They're not dead either, but they're not alive. If we go through the seven properties we outlined earlier:

1. Viruses don't really reproduce, they invade host cells and instruct the cell to replicate the virus.
2. Viruses don't maintain homeostasis.
3. Viruses don't grow.
4. Viruses use energy, but they do so vampirically when they latch onto a host cell.
5. We can't yet comment on whether they respond to stimuli. The jury is still out.
6. Viruses definitely mutate and evolve.
7. Viruses are made out of smaller building blocks (just not cells in this case).

So, there's only really two ticks, and maybe two half-ticks? By this definition, viruses are not a form of life, hence, not alive. Viruses just sort of drift around in a dead or dormant state until they can invade a living cell, hijack and reprogram it and all its resources, and replicate over and over. As a result, they're hard to combat because it's pretty tricky to kill something that's already dead, and then once it's latched onto a host cell, it's hard to kill without hurting the host's own cell! It can be quite the challenge, but one that medical science has tackled again and again.

Picture this: the Sun has formed, and all sorts of chaos has ensued as the disc of gas and dust swirls around the Sun and begins to coalesce into planets. First, the gas giants, Jupiter and Saturn, then the ice giants, Uranus and Neptune, and finally, the inner rocky planets. Earth formed and (as we discussed in Chapter 5) there was a great big impact as Theia smashes into the Earth (proto-Earth, technically speaking) and the Earth–Moon system formed. This puts us at roughly 4.5 billion years ago, give or take a few tens of millions of years.

What happened next? In the lead-up to this point, Earth had only recently formed; it had experienced an almighty collision with Theia so, unsurprisingly, it was all pretty hellish on Earth at this time. So hellish, in fact, that this period of geologic time is referred to as the *Hadean aeon* after Hades, the Greek god of the underworld. Not a pleasant time to be around at all. Fortunately for life, this period of time passed. Unfortunately, it took about half a billion years. But sometime afterwards, the origin of life, or *abiogenesis*, kicked off.

So what happened? If I knew the answer to that question I'd be writing a very different book!

It's one of the biggest mysteries in science. There are a few hypotheses, but the one currently leading is the *Heterotrophic theory*, or that of the *Primordial Soup*. The name refers to the build-up of various chemical molecules and materials required for biological processes (referred to as biogenic elements) in Earth's oceans, leading to a hot rich 'soup', within which life may have begun. The general idea is that the simpler, inorganic chemical molecules mixed and reacted to form more complex organic molecules and, eventually, became *biomolecules* such as proteins and nucleic acid. Some of the earliest forms of life sprang from these biomolecules.

Obviously, it's very hard to test the origins of life, but certain steps of this process can be examined in more detail. That is exactly what Stanley Miller and Harold Urey did in 1953 with the famous

Miller-Urey experiment. The Miller-Urey experiment worked by simulating the environment of the early Earth and seeing what might happen. To mimic the conditions of the early Earth, they combined water vapour, methane, ammonia and hydrogen in a sealed flask; then they fired electrical sparks (acting as a sort of lightning strike) into the flask. What they found was that, under these conditions, the basic chemicals from the early Earth reacted to form amino acids. More than 50 per cent of the amino acids found in life!

Ultimately, it turns out that the early Earth's atmosphere was likely quite different from what was originally thought, but that shouldn't detract from the Miller-Urey experiment's key takeaway: we can form complex organic molecules from simple inorganic chemistry. While the experiment doesn't answer the question, how did life begin?, it definitely helps us figure out one of the possible pathways, there's just still more nuance and detail for us to unpack first.

EVIDENCE OF ANCIENT LIFE

Even though we're not really sure how, life kicked off in some way or another. We do know that it was likely sometime in the next 0.5 billion years *after* the Hadean aeon, anywhere from about 3.5–4 billion years ago. Some of the oldest evidence of life is actually right here in Australia: there are *stromatolites* in the Pilbara of Western Australia that have been dated to be 3.48 billion years old. Stromatolites are rock structures produced by *cyanobacteria* (one of the earliest life forms) and are strong indicators of ancient life. Figure 6.1 shows what the stromatolites in Western Australia look like.

In 2016, however, Australia lost the 'Crucible of Life' crown, because similar evidence of ancient stromatolites was discovered in Greenland, dating back 3.7 billion years. Greenland didn't hold the title for long though because, in 2017, we discovered evidence of biological activity (read: life) in Quebec, Canada, dating all the way back to 3.77 billion years and possibly as far back as 4.28 billion years. (As you can imagine—there's some uncertainty when trying to look back that far!) There's not much room between 4.28 billion years ago and the formation of Earth, so it might be hard to topple that record, but we still might discover something older.

Figure 6.1: Stromatolites in Western Australia.

The evidence of ancient life discovered in Quebec is special, not just because it is the earliest known evidence of life but, perhaps even more importantly, because those ancient, fossilised structures are similar to microorganisms found today around hydrothermal vents. Before I race too far ahead, just a quick note on what hydrothermal vents are. We touched on them in Chapter 3, when we discussed how the Earth's core is home to violent geological activity. Things can get immensely hot and highly pressurised. This energy can often make it to Earth's crust and burst through the surface, either on dry land in the form of volcanoes, or underwater in the form of hydrothermal vents. In the latter case, the energy bubbles up through the vents (or cracks) in the ocean floor as hot, black smoke laced with sulfides or sulfur-bearing minerals.

Hydrothermal vents are sometimes referred to as *black smokers*. They are extraordinary because the energy from the Sun doesn't penetrate deep enough to reach the ocean floor. The environment deep underwater on the ocean floor is therefore missing one of the key ingredients for life: energy. That's where the hydrothermal vents step in. They spew out energy and water rich in dissolved minerals, resulting in a warm, wet and nutrient-rich environment. And wouldn't you know it, around this environment, we find life.

The most incredible thing about this life is that it's a wildly different ecosystem to what we see near Earth's surface, where the Sun is the energy source. Here, we find these unique life forms, tube-worms, which don't harness energy from the Sun, but harness it from the planet, and they do so much like plants using photosynthesis. Of course, different environment, different energy source, and so tube-worms use these sulfur compounds in an analogous process called *chemosynthesis*. So despite these stark differences, this environment still manages to support life.

PANSPERMIA

What if life didn't start here on Earth? What if it started in some distant, far-off location and just ended up here? That forms the basis of the *Panspermia* hypothesis: that simple microorganisms, or microbial life, are actually far more common throughout the Universe than we might think. Microbial life may end up hitching a ride on asteroids or debris flying through space, spreading further throughout the Universe. In such a scenario, one of these asteroids or some debris riddled with life collided with Earth sometime after it had formed, thus delivering life to begin its story here. Of course, we've found no evidence that life arrived in this way (because we've found no evidence of life other than what is here on Earth). The bigger problem with panspermia is that all it does is relocate the question of: How did life begin? from Earth to some other distant planet. The hypothesis doesn't answer the question at all, it just kicks the can down the road. It's a fun and out-there idea to think about, but doesn't really bring us any closer to an answer.

But I digress—again! Let's take a few (hundred) evolutionary steps backwards from the tubeworm to the microorganisms that appear to resemble the fossilised structures in Quebec. Why do we care so much that some ancient, fossilised life form might resemble micro-organisms found around present-day hydrothermal vents? Because it might give us clues as to what early life looked like on Earth. Maybe, this is where life started? Maybe, this is where *and how* life started? Having a living system today that resembles the conditions of how life started on Earth 4 billion years ago is remarkable, and it could help us unlock the secrets to life and answer some *very* big questions!

There's another reason it's exciting: maybe, this is what we should be looking for in our search for life beyond Earth? Let that idea percolate in your head, let your imagination run wild, and we'll return to this topic in much greater detail in Part 2.

THE GREAT OXYGENATION EVENT

This is a highly accelerated and shortened summary of life on Earth; we're simply focusing on the bits that are most relevant to what we're discussing in this book. With that in mind, the next thing I want to discuss is something called the *Great Oxygenation Event*. This is a pivotal moment in Earth's history and for life on Earth, and our good friends cyanobacteria (remember the architects of those ancient fossilised structures, stromatolites?) are responsible.

We know from Chapter 2 that our atmosphere is fundamental to life on Earth. We also hinted that, for us, oxygen in the atmosphere is really important. Remember respiration (the process for bringing oxygen inside our bodies)? We need to bring oxygen inside our bodies for a key property we highlighted earlier: metabolism. Oxygen is important for converting chemical energy into usable energy in our bodies. Has oxygen always been in the atmosphere, available for animals to use?

The short answer is no. For the first couple of billion years there was little to no oxygen in the atmosphere; it mostly consisted of nitrogen (N_2) and carbon dioxide (CO_2). As you may have already guessed from the name of it, that all changed during the Great Oxygenation Event. Essentially, cyanobacteria pumped out so much oxygen that it started to accumulate in the atmosphere. The problem with this theory is that cyanobacteria had been around for quite a while before the Great Oxygenation Event. So why was it that oxygen started to accumulate only at about 2.5 billion years ago?

Well, it's hard to say, but there are a few hypotheses. One is that, potentially, there were a lot of volatile chemicals in the ocean, such as carbon or iron, which were somehow buried. Oxygen reacts and bonds strongly to these elements, thus removing them from the atmosphere. If these elements were in fact buried, the oxygen would no longer have anything to bond to and would have begun to accumulate in the atmosphere. As we mentioned in Chapter 3, this situation could have been triggered by Earth's rocky surfboards—plate tectonics. Perhaps a large portion of iron-rich land was subducted and buried deep below the Earth's surface allowing oxygen to begin its accumulation?

Another hypothesis suggests the event was caused by the jump from single cellular to multicellular life. There's a body of research suggesting that cyanobacteria may have evolved at around this time to become multicellular, giving them an advantage in both their mobility and survivability. Perhaps this evolution to a superior variation allowed cyanobacteria to spread far more rapidly across the planet and produce enough oxygen so that it accumulated in the atmosphere. We still don't know exactly what happened during the Great Oxygenation Event, or why it happened, but we do know it's likely to be the single most important atmospheric event to occur on Earth.

THE CAMBRIAN EXPLOSION

So, now there's oxygen in the atmosphere. Earth is beginning to resemble something similar to what we know today, but it's not quite there yet. There was another remarkable transition that took place, which we briefly mentioned in Chapter 3: the *Cambrian Explosion*. Here, Cambrian refers to the Cambrian period—a period in Earth's geologic timeline similar to the Hadean aeon, although a smaller unit of time (an aeon is made up of eras, and an era is made up of periods). The Cambrian Explosion was an event that happened during the Cambrian period about half a billion years ago, when the diversity and variety of life increased dramatically.

Prior to the Cambrian Explosion, life forms were like our friends, cyanobacteria: fairly simple, single-cellular or small multicellular organisms. After the Cambrian Explosion, life not only diversified, but grew exponentially in complexity. Evolution proceeded at a rapid rate, and features that exist in modern animals began to form: legs, eyes, gills and jaws. But what caused this explosion? What caused this rapid change and diversification of life that had existed in simple form for so long?

We're still disentangling that question, too. There are a few hypotheses. One is that something drove another steep increase in oxygen in the atmosphere, allowing for larger and more complex organisms to develop. Another is that the evolution of certain abilities drove the explosion, specifically eyesight or vision. Perhaps the edge this gave predators to detect prey from afar drove the evolution of various defence mechanisms, which then led to better offence mechanisms, and so on. A sort of predator–prey arms race.

We're still unsure, and it's most likely not attributable to any single trigger but rather the combination of several triggers. Just as the Great Oxygenation Event is arguably the single most important atmospheric event to occur on Earth, the Cambrian Explosion is the single most important evolutionary event.

Once life diversified during the Cambrian Explosion, eventually, over hundreds of millions of years, it evolved into what we know today. From the early formation of Earth, we made it here: a complex, bustling world of varied environments and ecosystems, bursting with the rich diversity of life. It's hard to completely wrap our heads around the timeline so I've summarised the key steps in Figure 6.2.

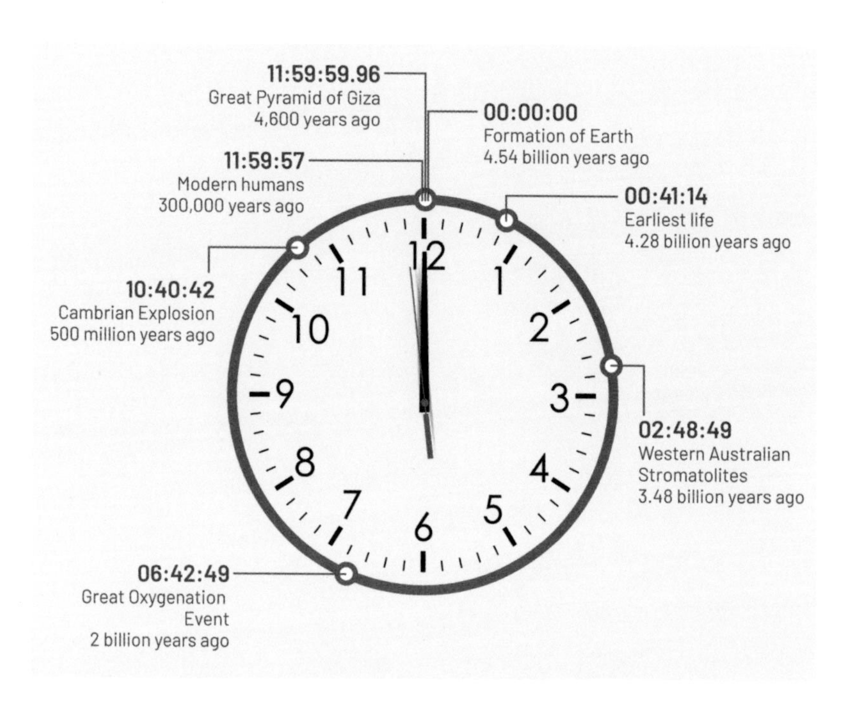

Figure 6.2: The timeline of Earth highlighting some of the key stages from creation through to the diverse life we see today.

To date, Earth is not just the only place we know where intelligent life exists, or complex life, or simple life—it's the one place where any form of life exists. What we really want to know is whether that's because it's the only place with life at all, or whether we just haven't found it yet. Do we simply need to look harder?

We'll get to that soon enough, but first let's consolidate things.

In the first part of this book, we focused on Earth and what makes it special. What is it about Earth that has made it the only place we know that supports life? We looked at:

- The Big Bang, and how water forms, supports life and allows for vital nutrients to be transported around the body.
- How Earth is enveloped in an atmosphere that lets us breathe, but also ensures water is usable in liquid form.

- Earth's magnetic field and the role it plays in protecting us from the flying fusion reactor in the sky (the Sun).
- Plate tectonics, and its impact on regulating the temperature of Earth and circulating chemicals and nutrients around different environments.
- The formation of the Moon, via the tremendous impact of Theia with proto-Earth, and how the Moon helps to support life on Earth.
- The beginnings of life on Earth, when life likely started and some of the key events from then to now, including the Great Oxygenation Event and Cambrian Explosion, both of which depended on several of the aforementioned properties of Earth.

Now, let's move further outwards: *Beyond Earth*. Where should we look harder? Where should we look to see if life has—or had—a chance at surviving on another world? Are there other places inside our own Solar System where we should look? Are there also other places beyond our own Solar System, worlds circling other stars, that we should examine?

Where else might be special?

PART TWO

WHERE ELSE IS SPECIAL?

We've established that Earth is quite a special place. It's almost limitless in its beauty and supports an astonishingly diverse array of environments. Whether in the permafrost-riddled tundra of the Alaskan wilderness, the tropical jungles of South-East Asia, the arid landscapes that sprawl across Australia, or even the unknown darkness deep below the ocean's surface; regardless of the environment in question, life thrives and blossoms. Some of this success is due to Earth's life-nurturing properties, but some may be attributed to our ingenuity as humans. As a species, we have sculpted, moulded and transformed almost every corner of the Earth, and we have manufactured the necessary attire and technology to let us live comfortably nearly everywhere on the planet. It's safe to say, Earth doesn't just support life, it fosters it.

Where else do we find such fortuitous circumstances? This question has played to our inquisitive nature for . . . well, likely from as early as we've had the cognitive capability to think about it. And it continues to stoke our curiosity. Before we start delving into the Pandora's box that is the search for Earth-like planets, we should consider this question from a different perspective.

Rather than considering a planet that would be habitable *to us*, we should look for a planet that is habitable *to life*. As exciting as it is to immediately lurch forwards towards the shiny object—turning over in our minds the idea of little green men, advanced civilisations and contact—we must walk before we can run. First, let's think about where simple life could have existed. Where could it still exist? Where could we find the same set of conditions that yield the right environment for life to begin, just as they did here, on Earth?

So, where else is special?

CHAPTER 7

REQUIREMENTS

Earth is a nifty little planet, isn't it? It's our shiny blue marble that every human who has ever lived calls home. In Part 1, we went through what makes Earth special—water, the atmosphere, plate tectonics, the Moon and our good friend Jupiter (or at least, our existence suggests they're our friend!). These properties are fairly important, if not critical, not only to kickstart life on Earth, but to provide the perfect environment for life to evolve and blossom. So now, let's move beyond Earth. Let's start thinking about where we should be looking to find other special places—other places where life might exist. But first, let's think about what it is we're looking for. Should we focus on everything we covered in Part 1? Should we prioritise some things over others and, if so, which? Is it safe to assume that we're looking for life like us?

Let's start right at the top. What are we even looking for?

Life, right? That'd make this whole 'Are we alone?' thing really easy. If we had technology at our disposal that we could just point at some other world and it would determine whether life exists, or has existed, we could wrap things up right there. That'd be Nirvana, sure, but it's also Dreamland. We don't have that sort of technology yet. Maybe someday we will, but not yet. Because we're constrained by technology, we've got to use our human ingenuity to engineer instruments that can detect things to help paint a picture, shed some light and provide us with some answers. If we're not looking for life, then what are we looking for?

WHERE CAN LIFE LIVE?

Typically, we're looking for places that are considered *habitable*. A few questions probably come to mind immediately. First, what does 'habitable' mean? On Earth, it generally refers to a dwelling that is suitable to live in. Your dwelling is considered habitable if it provides for your most essential human needs—shelter, safety, warmth, food and drinkable water—while also being clean. If we take a step back from a dwelling or housing context, and generalise things more, we can think about habitability in reference to places and locations being livable. Is the desert livable? The Arctic? The bottom of the ocean?

Your answers to these questions should point to the second question (which may have popped into your head already): Habitable for whom? Is the desert habitable? For us? Maybe not. For thorny devils? You bet. The Arctic? Not for us without a lot of technology and infrastructure. For polar bears? Absolutely. Bottom of the ocean? Not a chance for us humans. But it's home to a truly weird assortment of animals that dwell deep in the dark depths of the ocean: anglerfish, goblin shark, dumbo octopus.

The point is that whether somewhere is considered habitable depends on who we're considering to live there. So, if we're looking for somewhere that is habitable, are we looking for somewhere that is habitable for us? For thorny devils? For dumbo octopuses? The answer is complicated. We're certainly interested in places habitable for us, or nearly habitable, so we could engineer a way to make it workable for us to live, but why?

DON'T BE A CARBON CHAUVINIST!

When it comes to our search for life beyond Earth, there's something peculiar to think about. Science fiction often depicts aliens as being bipedal, if not humanoid. Think of the quintessential 'little green man', ET, Marvin the Martian, Xenomorphs (from *Alien*), Prawns (from *District 9*)—it's a common trope for aliens to be human-like.

There are obvious exceptions to this, which raise an important consideration: aliens won't necessarily resemble humans at all. When it comes to searching for life, there's also something more fundamental to consider than how accurate science fiction is. Will aliens, or life beyond Earth, be like life on Earth *at all?*

All life on Earth is made up of the element carbon. At a chemical level, carbon is so synonymous with life that a whole branch of chemistry is devoted to carbon compounds, called *organic chemistry*. Carbon is important to life here, but what if that's just something that happened on Earth by chance? Is there something intrinsic to carbon that makes it perfect for life, or could something else with similar properties be just as effective?

An inspiration of mine, astrophysicist and science communicator Carl Sagan, coined the term *'carbon chauvinism'* to refer to the idea that life could be radically different from how it exists here, down to the very building blocks of life itself. If that were the case, then it turns our search for life beyond Earth on its head! If not carbon, maybe silicon? Is it water we need, or will any liquid do? We'll explore this in more detail in Chapter 9.

Without diving too deeply into things, becoming interplanetary by colonising Mars (or another planet, for that matter) is a sort of *existential redundancy*. If we went the way of the dinosaurs and were made extinct by something out of our control (such as a large object colliding with Earth), having another colony on Mars would ensure our continued survival. Once the dust settled (this would take a long time!), we could recolonise and repopulate Earth.

This discussion about existential redundancy and colonising Mars is large and complicated enough to warrant its own book, so let's get back to 'habitable for whom?' For us, yes, but we're also keen on the simpler stuff, because discovering life—even simpler forms—helps us to gain an understanding about how life works, and starts to provide little nuggets of knowledge that contribute to answering the question: Are we alone? We're not just interested in finding places habitable for us, but also somewhere habitable for really simple life. That's why, in Part 1, we needed to lay down so comprehensively what it is that makes Earth special; what makes Earth habitable *not just to us* but to all life.

A SOURCE OF ENERGY

I'm going to be a little cheeky here, and first discuss something we *didn't* talk about explicitly in Part 1, but that's because it doesn't pertain exclusively to Earth. I'm talking about the Sun or, more generally, a source of energy.

We did touch on the Sun in Chapter 6, when we mentioned that life needs energy (for example, plants can convert sunlight into usable energy by photosynthesis). Life needs an energy source because several processes are *endothermic* (they require energy to occur). What this means is that certain processes require overcoming an *energy gradient*, which is sort of the equivalent to pushing something upstream. Imagine trying to blow air into a balloon that's already inflated. The balloon will naturally deflate (the air will escape) unless you expend energy (breathing more air into the balloon) to inflate it further. Life needs energy to overcome all sorts of processes like this at microscopic levels unseen to us, pushing and moving things around against the direction they naturally want to go (the sodium-potassium pump is one of the most well known). On Earth, the Sun is our primary energy source; typically, when we're looking for other places with an energy source (for example, planets), we're looking for bodies that also have a host or parent star, like the Sun.

For places within our Solar System, we already know there's a big fusion reactor in the middle—the Sun—and for places outside our Solar System, well, fortunately, stars are conveniently bright. They're some of the easier things to spot in space so, generally, this requirement is easily met because the way in which we search for planets is by looking around stars (more of this to come in Chapter 10). There are other energy sources though, which we'll get to in the coming chapters.

WATER, WATER, EVERYWHERE

It probably comes as no surprise that water is the biggest focus in our search for life and worlds we might deem to be habitable. We've established how important water is to life in Chapter 1, but I want to emphasise that water is so vital to life here on Earth, it's enough to overcome even the most extreme environments. What I mean is that, despite a particular environment here on Earth appearing to have conditions far too hazardous or extreme for life to exist, if we find liquid water there, then (almost without fail) we find life.

Extreme heat on Earth? No sweat. *Pyrolobus fumarii* (a type of microbe) is alive and kicking at 113°C. Extreme cold? Too easy. The

Himalayan midge is active at −16°C. What about pressure? At the bottom of the Mariana Trench, which is nearly 10,898 metres below the sea floor, the pressure is 1100 times as strong as at sea level. That's pretty hard to get your head around, so picture this: imagine you're lying down on the floor and someone draws an outline of yourself. For the sake of easy calculations, let's say that the area of your body inside that outline is 1 square metre. We can now calculate the amount of mass that would have to be on top of you to experience the pressure at the bottom of the Mariana Trench. I'm going to do this with people. We're going to pick an easy number for the weight of a person, let's say 100 kg (it doesn't matter if you weigh more or less than this; but 10 kg is too light and 1000 kg is too heavy, so 100 kg is a round number that's a good fit for our calculations).

How many people have to be lying down on top of you for you to feel the pressure at the bottom of the Mariana Trench? 10? 100? 1000? Way off, actually. It's closer to 100,000! That's right, lie down and stack the number of people in a full-capacity Melbourne Cricket Ground on top of you, and you'd be experiencing the pressure at the bottom of the Mariana Trench. It's definitely *not* habitable for us, but for the *Mariana snailfish* that lives at a depth of 8000 metres? You bet!

We keep finding life that survives in these crazy environments, over and over again. Overly acidic environments? No such thing for *Cyanidium caldarium* (a type of red algae). Overly salty? Not for *Dunaliella salina* (a type of green algae). Radiation? Not a drama in the world for *Deinococcus radiodurans* (a type of bacteria). As long as water is available, life has found a way to make it work.

The next logical question is this: How do we search for water? There are several ways we can determine if water exists somewhere, but let's focus on three specifically.

LITTLE WATER BEARS

While we're on the topic of weird and wonderful creatures, it would be remiss of me not to mention tardigrades, or *water bears* as they are sometimes nicknamed.

Tardigrades are funny little organisms that are best known for being particularly hardy. We don't classify them as extremophiles (organisms that live in extreme environments) like some of the exotic creatures we've mentioned in this chapter because they don't live or thrive in extreme conditions. What they are able to do, however, is endure and survive extremely hazardous environments in short bursts.

They've been known to survive extreme heat, extreme cold, dangerous radiation, crushing pressures and even survive the vacuum of space! They do this by entering a special state called *cryptobiosis* where the organism essentially halts all its metabolic processes until it's safe to fire them up again. They can sometimes stay in this state for as long as thirty years! Do these cute little water bears hold the secret to long-distance travel through space?

WE CAN SEE IT

The first way is the most obvious: *we can see it.* Ice water? We've found it on Mars. It exists at the Martian poles. We've snapped it; we've seen it. Liquid water? Seen that, too! Okay, I don't want to ruin the surprise (all will be revealed in Chapter 9) but we've taken photos of jets of water, right here in our very own Solar System, on an object that *isn't* Earth.

WE CAN DETECT IT

While seeing is believing, it's also critical to chemically verify that the ice (or liquid) we can see is indeed water and not something else. We must make follow-up observations to detect whether or not what we spotted is actually water. Conveniently, one of the key ways to verify the chemical composition of water we can see is also the second method by which we search for water: *we can detect it.* Because 'detect' is fairly all-encompassing, I should specify that I mean *detect using spectroscopic techniques.* This raises an important question: What is spectroscopy?

Spectroscopy is the study of how matter and light (or more generally, electromagnetic radiation) interact. *Spectro*— comes from 'spectrum', because it's the spectrum of light we're interested in. You may be familiar with the light spectrum from a simple science experiment, in which you shine a white light into a triangular prism and the light is refracted into a rainbow. Spectroscopy is tremendously important to almost all fields of astrophysics, shedding light (can't help making a good pun) on the properties of stars, planets and galaxies; the 'space' between stars (the interstellar medium); and the motion of objects in the Universe. The interactions between matter and light are an extraordinary property of nature, and observing these interactions through spectroscopy has a plethora of applications, not just in astrophysics but in other fields of science such as acoustics, materials science and medical science.

Spectroscopy can detect water via a technique called *absorption spectroscopy*, which works because of some fundamental properties of nature. Cast your mind back to Chapter 1, when we discussed electrons and likened them to eggs. Nature likes eggs to be in full egg cartons (that vary in size, depending on the element). It turns out the eggs (electrons) can also be in different states, which have different amounts of energy associated with them.

There's no neat analogy for this, so let's get creative: imagine our eggs are *dancing* eggs. In the lowest energy state, each egg is just standing there, stationary, not dancing. The next state is a little more energetic; the egg is now stepping from side to side—nothing too wild, but certainly getting into a groove. Moving on, the next state is really getting into it and ripping out the Macarena. The final state is really up and about; the egg's doing a full-blown, highly choreographed solo dance routine (let's call it the Egg Beater).

Just as nature likes full cartons of eggs (electrons), it also prefers its eggs to be relaxed and at rest: in that low-energy state of standing still. This is where light—and the energy that light carries—comes in. Light is special in that it has energy; the amount of energy light has is determined by the frequency of that light. While the frequency of light might sound like a weirdly abstract concept, fortunately, we have a highly sophisticated instrument for converting light frequency into something more easily understandable: the human eye. Our eyes can perceive the visible portion of the electromagnetic spectrum and differentiate (except in the case of colour blindness) these frequencies into different colours. In this way, we learn that low-frequency (and hence low-energy) light corresponds with the red end of the rainbow, while high-frequency (and hence high-energy) light corresponds with the blue end of the rainbow.

Okay, back to our dancing eggs! How do our eggs start dancing? Basically, the energy needed to get the eggs from standing still to stepping side to side will be equal to a specific amount of energy, which will correspond with a specific colour. As mentioned, red light has less energy than blue light. If we shine a particular shade of red light on our eggs—BAM!—our eggs start stepping from side to side. Imagine if, instead of a red light, we used a more energetic colour, say, yellow light? That might take our eggs from standing still to doing the Macarena (it all depends on what nature has decided is the right amount of energy for the eggs to dance the Macarena). Maybe, when highly energetic blue light hits our eggs, they go from standing still all the way to doing the Egg Beater.

This might seem a bit silly, but it's important to understand that this is sort of how absorption spectroscopy works. Different chemicals absorb different colours of light because the energy of certain colours is equal to the amount needed to get that specific chemical's eggs dancing different dances. And it's important to note that these amounts are both very specific, and unique to the chemical. If we had millions of billions of these eggs and shone white light at them—which is made up of the entire rainbow (remember our triangular prism?)—then some eggs are going to absorb the red light to make them side step, some eggs are going to absorb the yellow light to make them do the Macarena and some eggs are going to absorb the blue light to make them do the Egg Beater. If we look at the white light on the other side of the cloud of billions of eggs, then we'd notice there's no red, yellow or blue light. It all got absorbed. This is why it's called *absorption spectroscopy*.

By looking at white light after it has interacted with something, we can use the missing colours to identify what that something was made from. The amount of energy and type of dance the eggs do will change, depending on which chemicals are present. Different chemicals absorb their own unique colours; it's almost like a signature or fingerprint.

How does this help us detect water? Because the water molecule—

H_2O—has a set of dances unique to it, and thus will absorb a unique set of colours of the rainbow. By looking at which colours are missing (which colours have been absorbed), we can confirm whether or not water is present. So, when we see things that look like water, we can use absorption spectroscopy to examine if the colours of water are missing—whether or not we can see water's chemical fingerprint. We can also use this technique when the thing we're looking at is so far away that we *can't see water.* I'm referring to places outside our Solar System. We'll dive into this in more detail in Chapter 10 but, basically, spectroscopy is useful regardless of whether we can actually see water with our eyes.

WE CAN HYPOTHESISE

The third way we can find water beyond Earth is by hypothesising where it could be. Remember our campfire star analogy from Chapter 1? Water evaporates if it's too close to the campfire and freezes if it's too far away, but between these two extremes is a sweet spot: an area where liquid water can exist. It's not too hot, not too cold, but just right. This area around a star, where the conditions are just right for liquid water to exist, is referred to as the *Habitable Zone.* If your first thought was porridge and a family of bears, you're not alone; it's also called the *Goldilocks Zone.*

Figure 7.1 shows what this looks like; it exists as a sort of doughnut around a star, a belt within which the conditions mean water can exist as a liquid. We use the Habitable Zone to help determine if a planet could have water on its surface. From the brightness of a star, we can determine how hot it is, and then from its temperature, we can work out where liquid water could exist—how near and far from the star liquid water could exist on the surface of a planet like Earth.

Within our own Solar System, we can use several other techniques to search for water (as outlined earlier), but the Habitable

Zone is extremely useful when assessing planets discovered outside the Solar System. In case your curiosity is getting the better of you, in our own Solar System, both Earth and Mars reside in the Sun's Habitable Zone. Does that mean Mars is habitable? We'll dive into this further in Chapter 8.

Figure 7.1: The Habitable Zone in our Solar System—a doughnut-shape, or annulus, that exists around the Sun.

PROTECTIVE SLEEVES AND SHELLS

So what else are we interested in? How about atmospheres? As far as habitability goes for us, it's a resounding yes, but what about habitability for life more generally? Let's continue along the path we have now well and truly established (that water is essential to life) to guide our interest in atmospheres.

In Chapter 2, you may recall that Earth's atmosphere acts as a sort of sleeve for water, creating a barrier to keep the water from boiling off. We're most interested in this mechanic—the sleeve—when looking for life. This expands our search, so we're not only interested in those worlds where an atmosphere can protect water from the vacuum of space, but also those where *any* protective shell can. A shell made, for example, from solid ice. We want to find liquid water, and that needs some means by which to prevent water from boiling off into space. Whether it's an atmosphere, or icy shell, doesn't matter as far as our search for life is concerned.

Of all the properties that make Earth special, we are most interested in finding liquid water in our search for life. Going through the list of properties in Part 1, we find that an atmosphere would be great, but we can also settle for an icy shell. How about the other properties? Tectonic plates? Maybe. We can search objects in our own Solar System for evidence of tectonic plates, but what about objects outside our own Solar System? Not a chance.

One property we do think we need—and which plate tectonics requires—is that the world is solid or rocky. Specifically, we want to find small worlds like the inner-most rocky planets Mercury, Venus, Earth and Mars, or the moons in our Solar System. We're not interested in places like the gas giants Saturn or Jupiter, or the ice giants Uranus or Neptune. These planets have fundamentally different structures, so much so that it is simply too hard for us to hypothesise how life might begin (and continue to exist) in such an environment.

That isn't to say life couldn't begin there, or that it doesn't exist, just that the conditions are so dissimilar to Earth's that we can't draw any comparisons. Without flipping our understanding of life on its head, we focus on rocky objects because, as we've mentioned before, there's only one place we've found life, and that's right here, on a rocky planet.

When it comes to searching for habitable worlds, the key requirements we're looking for are liquid water, encased in an icy shell or atmosphere, and located on a rocky planet. We may want to consider some other things if we narrow our focus to places habitable *for us*—for example, we definitely want an atmosphere to hold the liquid water down, not an icy shell—but for now we're thinking about searching for life in general.

With that in mind: Where else is special? As this chapter shows, places within the Solar System are much easier for us to investigate, observe and analyse in our search for life. So let's start by having a look at the interesting places in our own Solar System.

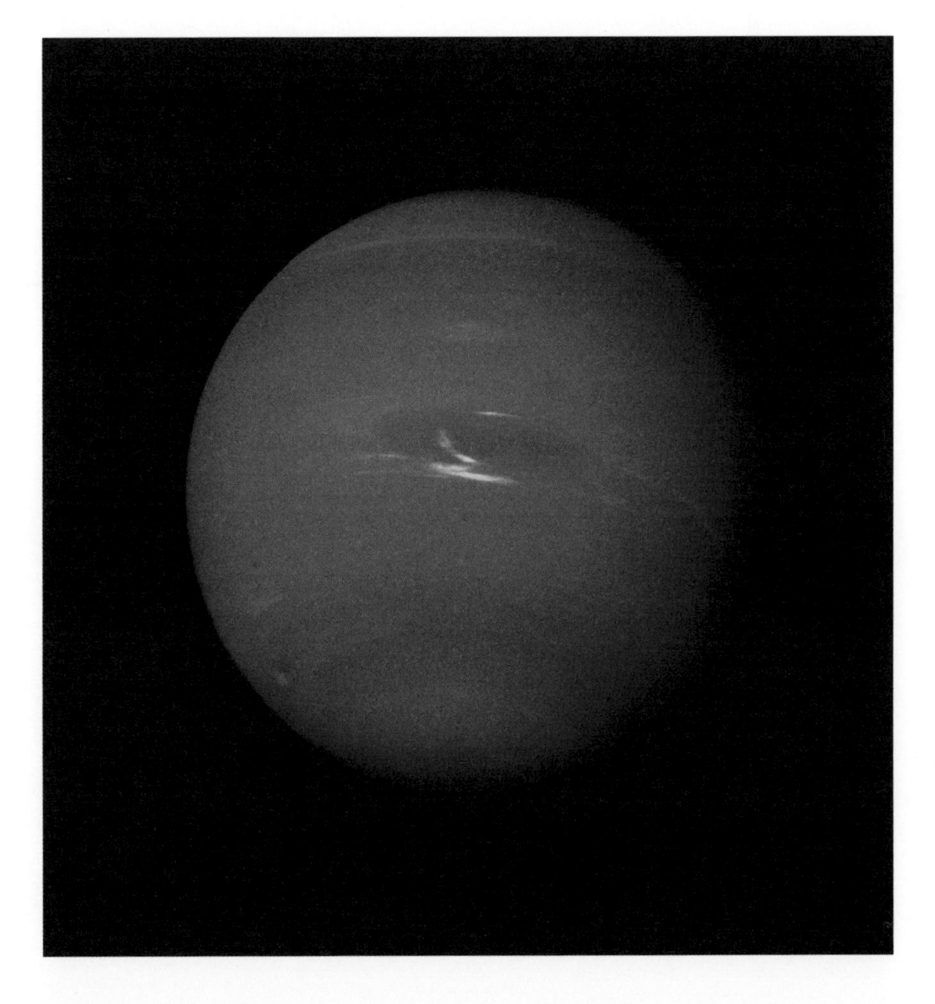

CHAPTER 8

MARS

When searching for life, we don't want to turn our understanding of life as we know it on its head: we want to keep the fundamentals there; we want to be looking for liquid water; but that's about as far as we want to limit our search. Naturally following on from this requirement is that this liquid water is being 'held down' to the rocky body somehow, either by an icy shell or an atmosphere exerting pressure. So what does that mean in the context of our own Solar System?

As we touched on in Chapter 7, the gas giants Jupiter and Saturn, or ice giants Uranus and Neptune, create challenging problems regarding the physical state in which water exists. Under the enormous gravitational pressure that exists on these massive planets, all sorts of exotic environments exist: super-compressed water, deep within the cores of giant planets; solid mixtures of water, methane and ammonia ices; and the planets' outer atmospheres, where various chemicals dance about in swirling clouds and immense storms.

The giant planets are beautiful, to be sure, but they're so fundamentally different from Earth that thinking about how life might exist there is a bridge too far for us. As such, we're mostly interested in small rocky planets, but when we're looking at objects within our own Solar System, we don't need to restrict our search to those within the Habitable Zone. We want them to be able to sustain liquid water in some capacity and we can observe objects in the Solar System with such fidelity that we can determine if they have liquid water, regardless of whether or not they're in the Habitable Zone.

In our Solar System, where can we find small bodies that could support liquid water? Well, there are only four small planets. We live on one of them, and another two are particularly hostile towards the existence of liquid water.

Mercury has no atmosphere and is far too close to the Sun, while the atmosphere on Venus is so thick, and has created such a powerful greenhouse effect, that the temperature is pumped up to more than 460°C, far too hot for liquid water to exist—but we'll get to that shortly. This leaves us with the one that has our robots rolling all over the surface. That's right, our friendly little red neighbour: Mars.

MARS *PATHFINDER* AND FINDING YOUR PATH

The Mars *Pathfinder* mission was a spacecraft launched by NASA in 1996 with the goal of deploying a small robotic rover, *Sojourner*, on the planet to conduct a variety of scientific experiments. *Sojourner* was the first rover to be deployed successfully on another planet besides Earth (although, earlier, some had been deployed on the Moon). *Sojourner* gathered incredible data through its onboard instruments, captured tens of thousands of incredible images from the surface of Mars and, most importantly, paved the way for all future rovers. *Sojourner* ushered in a new era of planetary exploration.

While *Sojourner* only traversed roughly 100 metres across the red planet's surface, its successors *Spirit* and *Opportunity* achieved an impressive 7.7 km and 45.16 km, respectively (*Opportunity* managed to operate for another 8 years after *Spirit* ran out of steam). Their successor, the behemoth *Curiosity,* has only traversed 25.23 km so far, although it's still up and running; because it's packed to the gills with instruments, it does far less travel than the older rovers; it needs to be conducting experiments like the responsible scholarly robot it is!

The two newest rovers, NASA's *Perseverance* and China's *Zhurong* (the first successful interplanetary rover launched by another nation) are up there now, too. Who knows how long they'll kick (sorry, roll) around up there?

The *Pathfinder* mission has a special spot in my heart, because it is my answer to the question I often get: 'What sparked your interest in space?'. As young boys and girls, I think we're almost universally enamoured by space, aliens and other worlds. It's a natural child-like curiosity. For me, the point where it turned into something more real and a little less abstract was the Mars *Pathfinder* mission. My mum showed it to me (thanks, Mum!) and I realised this stuff was actually happening. We were putting robots on other planets! I was hooked, and the various twists and turns in my life have, eventually, led me to writing this book. My hope is that, for some of you, this book is your turning point, your own *Pathfinder*.

Mars is our second-closest neighbour; it's only slightly further away from Earth than Venus is. Mars is a far less hostile environment than Venus for us to explore, so it has been the subject of far more rigorous observation and exploration.

What makes Venus so hostile? Venus has a surface pressure more than 90 times greater than Earth's, which has resulted in landers only surviving on the planet briefly before being crushed by the enormous pressure. Even more dangerous than the crushing pressure is that the thick, dense atmosphere has created a powerful greenhouse effect, which is dominated by carbon dioxide, a strong greenhouse gas. We discussed the greenhouse effect briefly in Chapter 3—these gases behave like the glass in a greenhouse, allowing heat in but preventing it from escaping—and this is incredibly strong on Venus. It's so strong that, even though Mercury is closer to the Sun than Venus is, the surface of Venus is hotter due to the power of its greenhouse effect.

In contrast to the hellish environment that Venus serves up, the atmosphere on Mars is far less hostile, although it's still far from habitable. The surface pressure on Mars is only about 1/100th that of Earth and is composed primarily of carbon dioxide. It's certainly not breathable air, but at least it's not crushing us to death or turbo-heating the planet to hundreds of degrees.

Because Mars is relatively close to Earth and has a *comparatively* (to the Venusian hellscape) welcoming environment, we have taken it upon ourselves as a species to populate the planet with robots. Granted, these robots only operate temporarily, face significant communication challenges, have limited motion and actions, and typically involve considerable human involvement, *but* they're still there. And every one of them has been an incredible achievement for science, engineering and humanity. The first autonomous helicopter—*Ingenuity*—took flight on Mars in April 2021, a huge accomplishment that heralds in the next generation of space and interplanetary exploration.

So, Mars is relatively close to us and relatively friendly, which has led to it being a key focus of our exploration efforts. What have we discovered about Mars? There are several exciting discoveries, which we'll go through now.

WHAT HAVE WE DISCOVERED?

The first exciting Mars discovery relates to one of the requirements we discussed in Chapter 7, about the Habitable or Goldilocks Zone. The Habitable Zone is a doughnut-shaped area around the Sun that indicates where liquid water could exist, considering only temperature (there are many nuances and caveats to the Habitable Zone but, for our purposes, this definition is sufficient). What you may have noticed from Figure 7.1 in Chapter 7 is that Mars sits comfortably inside the Habitable Zone. Mars also receives enough energy so that liquid water can exist on the surface. The problem with Mars is what we've discussed already: its atmospheric pressure is only about 1% of that on Earth. We can figure out what this means using the phase diagram of water on Mars, similar to the phase diagrams we looked at in Chapter 2. If you recall, on Earth the phase diagram looks like this:

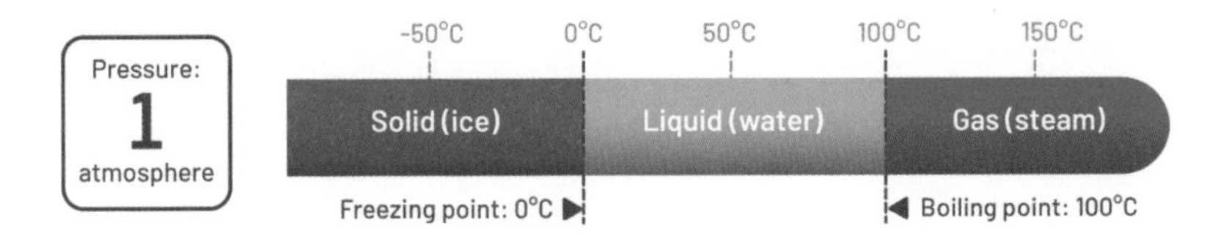

A familiar situation. Water freezes at 0°C and boils at 100°C. Great! But what happens on Mars? The atmosphere on Mars is just 0.6% of the pressure of Earth's atmosphere. In this scenario, the phase diagram looks like this:

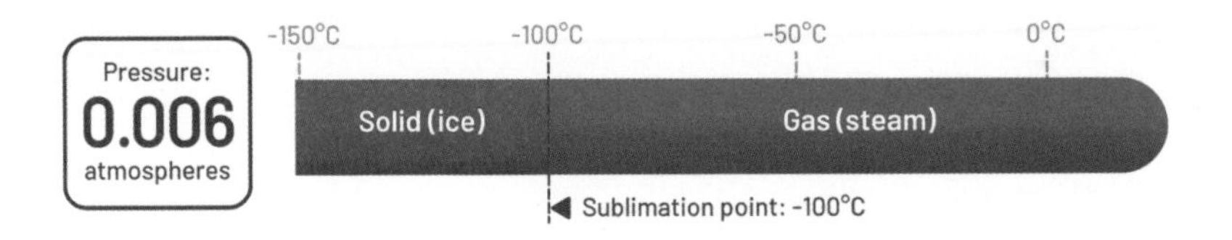

You can see the problem Mars faces having such a thin atmosphere. It's so thin that the atmospheric pressure is below the triple point of water (we discussed the triple point briefly in Chapter 1—that sweet set of conditions where water can exist as a solid, liquid or gas); as a result, water ice skips the liquid phase and sublimates directly into a gas. So, if it's cold on Mars, water exists as solid ice, and if it's hot (let's be honest—warm), then ice will just turn straight into gas. That's a shame because we really want liquid water; although water can still exist on Mars, it's in the wrong phase. What phase should we expect then? Being further away from Earth, we would expect Mars

to be cold, but how cold? With only a thin atmosphere, Mars doesn't have an effective greenhouse effect like that on Earth (or incredibly strong like that on Venus). This means the average surface temperature is about –60°C (although, in the warmer Martian summer around the planet's equator, it can reach a much more pleasant 20°C). Sitting at –60°C on average, we're well and truly below the sublimation point and would expect any water to be solid. And lo and behold, that's exactly what we see at the Martian north pole. Check it out in Figure 8.1. The clearest display of water on Mars is at the north polar cap—a beautiful (and quite abundant) white ornament, sitting atop Mars's crown. We're halfway there! There's actually a load more ice just below the surface of Mars, too, we just needed a little more investigation to find it, using spacecraft we've put in orbit around Mars (specifically, the *Mars*

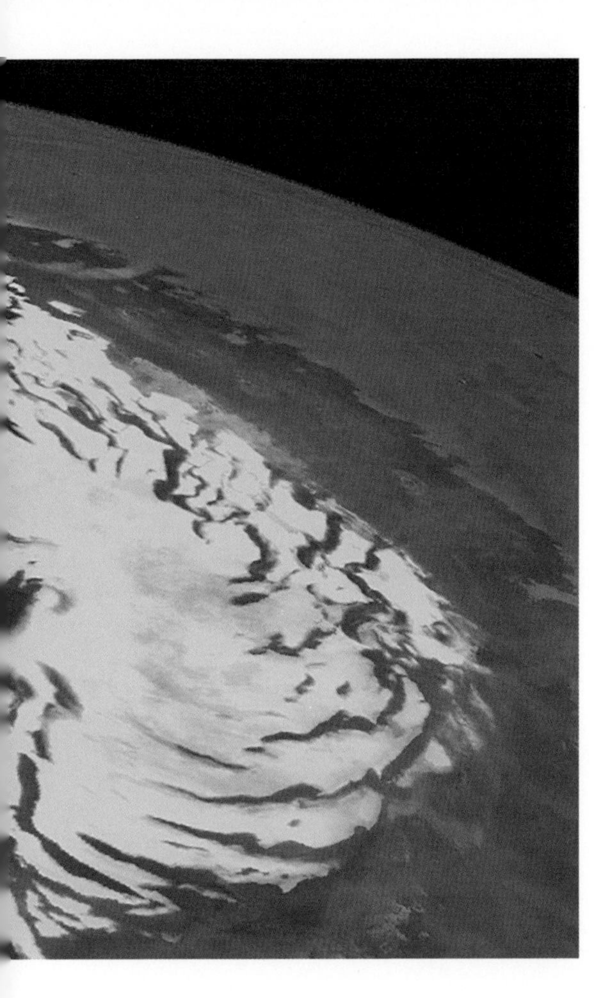

Figure 8.1: This beautiful imagery, captured by NASA's *Mars Global Surveyor,* shows the magnificence of the Martian north pole.

Reconnaissance Orbiter). In total, the amount of water on Mars is estimated to be about 5 million cubic km, enough to cover the entire planet in about 35 metres of water (about as deep as a 10-storey building).

ONCE A WATERY WORLD

Okay, so we've got ice on Mars. Water in a solid state. That's all well and good, but can we do better than that? Can we find some sort of evidence of liquid water? The phase diagram of water on Mars paints a pretty grim picture, but what if, instead of thinking about Mars as it exists today, we consider Mars as it used to be? From our comprehensive analysis of Earth and what makes it special in Part 1, it's increasingly apparent that our home is very much *alive*.

Earth is a big, dynamic changing system, which is true of many planets when they form. The planet formation process involves collisions and impacts, compression and crushing of matter under gravity, sliding and grinding of different objects—all processes that are naturally highly energetic and hot. What this means is that, when a planet forms, it can be chock-full of internal energy. This energy will slowly dissipate over hundreds of millions to billions of years and, in that time, the planet will be experiencing change. It will be evolving.

At the same time, the planet can be experiencing additional impacts and bombardments from other smaller bodies (which can deliver water or other elements, or create impact craters kicking up loads of matter into the atmosphere); blasts from highly energetic ions and radiation from the Sun, which can ionise and remove chemicals from the atmosphere; and even the atmosphere just slowly drifting off and evaporating into space. After a planet has formed, there's a period of significant evolution during which it can change dramatically. What this means with respect to Mars is that Mars as we see it today is not how Mars has always been.

Observations of Mars have provided us with some incredible imagery and rich scientific data, and when taken together and analysed thoroughly, it all points to something amazing. In the early 2000s, research began to show increasingly stronger evidence of ancient rivers, deltas and lake beds on Mars. While this liquid water has long since disappeared, the leftover geographic features of these ancient water systems are right there for us to observe. As both the number and sophistication of the vehicles we sent to Mars to take observations have increased, it's become clearer that Mars used to harbour significant amounts of water. In 2012, the *Curiosity* rover finally confirmed what all this scientific evidence had thus far inferred: *Curiosity* found direct evidence of an ancient riverbed in one of Mars's craters (the Gale Crater).

The enormity of this discovery might not be immediately apparent, so let's break it down. Ancient Mars harboured—for an extended period of time, I might add—significant bodies of flowing, liquid water. We know it was an extended period of time (think hundreds of thousands of years) because erosion is not a fast process: these valleys and deltas were carved out over millennia. We also know that the water was flowing because again, the erosion of the valleys and other geographic features like those observed requires a fluid to flow; for there to be motion. Finally, we know these bodies of water were large purely from the size of the features that remain on present-day Mars. Estimates put the largest sea on Mars, Eridania Sea, as having a volume nine times greater than all of North America's Great Lakes combined (nearly a quarter of a million cubic km!).

What this amounts to is that we have direct evidence that Mars used to have several environmental features that mimic what we have here on Earth. If life just requires a special collection of ingredients and the right set of circumstances, they might have all been present on ancient Mars. This is exciting because, while the conditions are completely different now and not conducive to life, it may have been in these bodies of water in the past. And that means that any remains of life if it existed on Mars would be evident in these dried-up ancient water beds!

So, not only do we have evidence that Mars had a good environment for life, but we know exactly where to look to see if life did pop up. That's awesome! Even more interesting is what happens when we look there. The discovery or lack of discovery of ancient fossilised life are both great results. We either find out that life simply needs the right ingredients and conditions, and that, with enough time, it'll spring up and (if the conditions remain just right) thrive, *or* we find out that there's still a missing piece to the puzzle. That what took place on Earth was special and we simply need to keep trying to discover what that is. Both results are exciting, and a little frightening, to think about.

We've shown that there's solid water on Mars *today*. And we've also shown that there was liquid water on Mars a long time in the past. Wouldn't it be great if there was liquid water on Mars today? Obviously, we can't have liquid water (or if we do, it will be very short lived—see the Mars phase diagram), but without spoiling things too much, we're getting closer.

SALTY WATER FLOWING ON MARS?

In 2015, an article was published that accompanied an announcement from NASA about discovering some evidence for hydrated salts in a feature that had been observed on Mars for years, the so-called *recurring slope lineae* (RSL). RSLs are narrow dark markings that appear in a cyclical, seasonal fashion on certain slopes of Mars during the Martian summer months, and which fade in the cooler months. This paper presented evidence strongly indicating that these RSLs were hydrated and briny (read: salty) chemicals! If something is hydrated, it means it's gained water molecules. Now, the evidence was not saying that liquid water was hydrating these salts, but that liquid water would be consistent with what was being observed.

The evidence supports the hypothesis that RSLs were salty, briny water, warming up in the balmy Martian summer and flowing down these slopes. Liquid water. On Mars. Flowing. Today! It's certainly a mind-blowing idea. If there's salty liquid water flowing on Mars, then that could be in biologically significant quantities, and possibly sufficient for life to be kicking about in.

We should note here that briny or salty water has modified chemical properties compared to pure water, and will thus have a different phase diagram to the one we looked at earlier (getting around our 'solid or gas only' problem).

Of course, science is rigorous, comprehensive and evolving. It doesn't stand still and constantly seeks to better understand and explain the world around us. Years later, further observations found a better explanation for what we were seeing; it turns out that grains of sand and dust slipping down the slopes formed a stronger hypothesis (and hydration being detected was because of these grains of sand and dust hydrating from the water vapour in the atmosphere—remember, 'hydrating' just means the water molecule is there).

This isn't an error of science, or a disappointing result, ruining our fun. This result still sheds more light on understanding the Universe around us. In this case, RSLs didn't turn out to be liquid water on the surface of Mars, but is that our only liquid water lead?

PROSPECTING FOR WATER

As of 2020, the answer is a resounding No! Science never rests, and in another tremendously exciting discovery, liquid water was detected on Mars. There's a caveat unfortunately (in fact, considering how much we've focused on needing a 'sleeve' such as an icy shell or atmosphere, you may have guessed it already). The liquid water is in the form of a subsurface lake.

Let's break down how this discovery came about. The European Space Agency's (ESA's) *Mars Express Orbiter* used an advanced radar instrument on board, which is capable of using radio waves to image buried geological features on a planet. This is similar to how seismic prospecting works on Earth, where sound waves are sent through the ground (often by firing explosives). How the sound waves travel and bounce around underground is detected by precise instruments and interpreted to describe the underground structures that are unseen by our eyes (a bit like a bat using echolocation in a cave).

Radio waves are good at penetrating through ice because there is little interaction, which makes them ideal for conducting this sort of underground prospecting through ice sheets, such as Mars's polar caps. The *Mars Express Orbiter* sent radar pulses towards Mars's polar caps. When these radar pulses bounce back, the underground structure can be mapped, based on how the signal has been influenced. Different materials will impact the signal (both its strength and how long it takes to return). Using this approach provided strong evidence to support our earlier hypotheses (remember the *Mars Reconnaissance Orbiter* from earlier in this chapter?) that there may be underground bodies of liquid water *through a completely independent technique!* That second point cannot be overstated.

In science, it's difficult to prove something because only one result is needed to contradict a hypothesis (consider the classic 'all swans are white' narrative, which was understood and accepted in Europe for centuries, until black swans were discovered in Australia and proved that statement to be categorically false). The equivalent scientific way to 'prove' a hypothesis is to be able to reproduce observations. Each time you're able to do so lends credence to the hypothesis you are putting forward. Using independent techniques to corroborate the same hypothesis is a strong way to support that hypothesis and so, as it stands, this is really strong evidence to support the notion of a subsurface liquid ocean on Mars as it exists today.

We've covered a lot in this chapter, so let's recap here. Mars appears to have a lot going for it, when we compare it with Earth. It's rocky like Earth, it's got an atmosphere like Earth (albeit, it's very thin and has different chemicals), but the big question we always fall back on with respect to looking for places that could support life is: Where's the water?

Well, we've shown that, in the past, ancient Mars had loads of water. So much so, it flowed across the surface, carving out valleys, channels and deltas, and leaving behind ancient riverbeds and lakebeds. These are perfect places to explore to see if life may have existed on Mars and help us to understand better if life really is just a simple recipe involving the right ingredients, or if something more exceptional took place on Earth.

Forgetting the past, though, the present is almost as exciting. We've got tonnes of solid water on the surface of Mars. It's still right there; we can see it. In terms of being somewhere habitable for us, Mars certainly ticks the box of possessing large quantities of the most essential ingredient for life. We just need to thaw it out under the right conditions and away we go. Even more exciting is that, right now (as of writing this in 2022), there is strong evidence to suggest the presence of subsurface liquid lakes. Right now. On Mars. Not only is there liquid water for us to filter and use, should we make the journey to set up a continuous or semi-permanent settlement on Mars, but incredibly, there could be life swimming about right there in those lakes, right now.

Mars has several places where water has existed, or still exists; we just need to look closely enough to find out whether or not life did, or does, exist there. Our little red neighbour is an excellent candidate for humankind's first interplanetary visit but, in the meantime, it's fortuitous that we can continue to send our little robotic friends to keep exploring on our behalf.

You most likely expected us to discuss Mars when talking about water and life but, as it turns out, Mars isn't the only place of interest in the Solar System. We've discussed that the planets closer to the Sun than the Earth—Mercury and Venus—are too hostile, which only leaves one option left.

Let's look a bit further out.

CHAPTER 9

SATELLITES

In the last chapter, we discussed the four terrestrial (or rocky) planets in our Solar System.

There is our home, Earth, the two hostile worlds of Mercury and Venus, and our friendly red neighbour, Mars. We also mentioned that, without a huge upheaval in our understanding of life, we can't really investigate the gas and ice giants of Jupiter, Saturn, Uranus and Neptune for life, because they're so fundamentally different from Earth. We need, instead, to focus on rocky bodies. So, if we've already considered the four rocky planets for life, what's left? Fortunately, there are several more bodies where liquid water could exist: natural satellites, or moons.

Earth's moon, aka *Luna* or simply *the Moon*, is a fairly typical-looking rocky object. It's large enough to assume hydrostatic equilibrium (scientific jargon for 'it's roundish') and, because it's rather small, it doesn't have an atmosphere. Remember, atmosphere is integral to maintaining liquid water on the surface of a body because it acts like a sleeve, trapping the water on the object. Without an atmosphere, water just boils off and drifts away into space. While our Moon is uninteresting from a liquid water standpoint, some of the moons orbiting other planets are riddled with intrigue. We'll discuss some of them in this chapter.

ENCELADUS

The first moon I'd like to focus on is one of the moons of Saturn, called Enceladus.

Enceladus is much smaller than our own moon, but it is large enough to be near-spherical (it maintains hydrostatic equilibrium). In Chapter 8, we found out that Mars is –60°C on average. As you'd expect, the further from the Sun we go, the colder we're going to get. As one of the moons of Saturn, Enceladus is roughly ten times as far from the Sun as the Earth is. And as you'd imagine, this means Enceladus is cold. *Really cold.* Measurements put it at about –200°C. It's a great icy white marble, hurtling around Saturn: the jewel in the Solar System's crown. Why, then, would we be interested in such a place? The answer comes from several scientific discoveries found over the course of the Cassini-Huygens mission to Saturn, a collaborative research mission among NASA, the European Space Agency (ESA) and the Italian Space Agency (ASI).

Cassini-Huygens launched on 15 October 1997 to study Saturn and its extensive system of rings and moons. Several spacecrafts had visited Saturn before: the *Pioneer 11* flyby in September 1979, and the *Voyager 1* and *Voyager 2* flybys in November 1980 and August 1981, respectively. Cassini-Huygens was the first mission that planned to enter the orbit of Saturn, rather than just performing a flyby. Even more ambitiously, the mission aimed to land a probe on the surface of Titan, one of Saturn's moons. The compound naming of the mission reflected these two key goals: the *Cassini* orbiter, designed to orbit Saturn, and the *Huygens* probe, designed to land on Titan.

GRAVITATIONAL SLINGSHOT!

One of the most incredible aspects of sending objects to planets and other bodies far away is the path a spacecraft takes from Earth to its destination. There's a really cool manoeuvre called a *gravity assist*; which is used to help propel a spacecraft along its path much faster and with less fuel. So how does it work? We could write an entire book on the wonders of space exploration, with a chapter on gravity assists; instead we'll just set the scene: *Dr Matt's Crash Course in Gravity Assists*.

Essentially, a gravity assist works by using another planet as a sort of slingshot but, instead of a giant rubber band storing elastic potential energy to propel something, we're using the giant mass of the planet and accompanying gravitational potential energy to propel our spacecraft.

Let's try another analogy to understand how this works. Think about lobbing a tennis ball towards a wall: you throw the ball, it makes contact and it bounces back to you at about the same speed (really, it'll be a bit slower because of friction and air resistance, but let's imagine a perfect world). This is the same as when our spacecraft encounters a planet *if the planet wasn't moving*; our spacecraft enters the planet's gravitational field, slings around the planet and flies off in a different direction with the same velocity at which it approached the planet. Great.

Of course, we know planets aren't standing still. They're hurtling through space in orbit around the Sun. So let's go back to our tennis ball analogy; but this time, you're standing on a train platform and lobbing the tennis ball towards the front of a train as it approaches. You throw the ball, it makes contact with the front of the train and it bounces back off at a far greater speed than you threw it. This is like what's happening with a gravity assist: our spacecraft encounters a planet hurtling around the Sun, it slings around the planet and flies off in a different direction, but this time, it leaves much faster than when it arrived. The planet's gravity and motion have given our spacecraft a boost in the direction it was going.

Cassini-Huygens utilised four of these gravitational slingshots on its journey to Saturn.

After leaving Earth, it swung by Venus then, after almost an entire orbit, it swung by Venus again and then, funnily enough, it swung back around Earth, picking up speed to fly by Jupiter before eventually arriving at Saturn, after a journey of almost seven years. At the start of July 2004, *Cassini* carried out its *insertion burn manoeuvre*. (This is a special operation used in spaceflight to shed excess velocity from the spacecraft. Remember our gravitational slingshot? Unless the spacecraft slows down enough to stay in orbit, it will just fly in, swing by and fly off from Saturn. We want it to slow down enough that it sticks around like a moon orbits a planet.) It began orbiting Saturn and, nearly six months later, in late December, the *Huygens* probe separated to begin its journey towards Titan. The complexity, distances and precision of these space exploration missions still blow my mind.

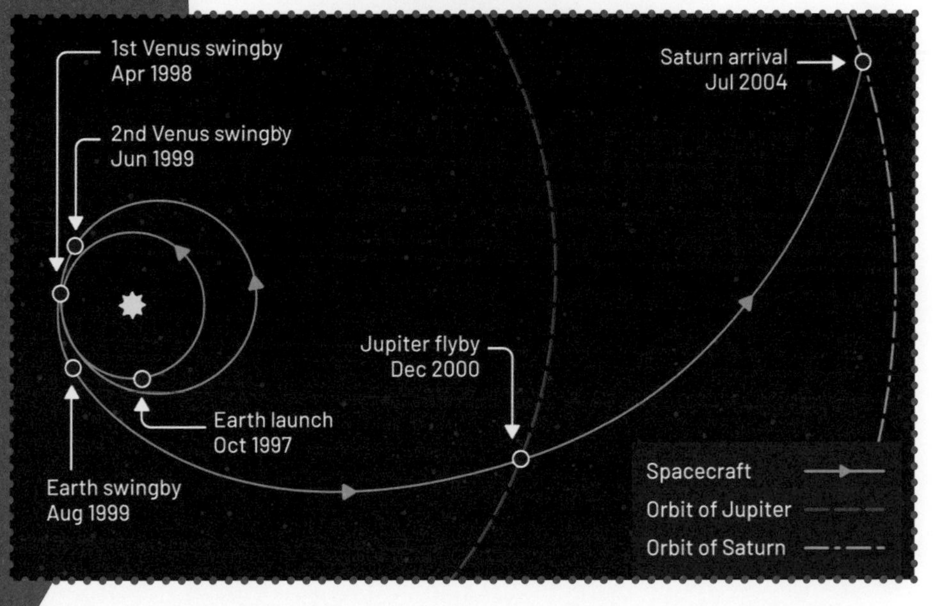

Figure 9.1 The trajectory of *Cassini-Huygens* from Earth to Saturn. The mission launched in late 1997 and passed by several planets to utilise gravity assists. Adapted from: Matson et al. 2002.[3]

Now this might come as a surprise—weren't we interested in one of Saturn's other moons, Enceladus? You're not mistaken. However, while the *Huygens* probe was on its way to land on Titan in an effort to uncover the secrets of its atmosphere and obtain out-of-this-world imagery of its surface, the *Cassini* spacecraft was in orbit around Saturn, meaning it could carry out some wonderful science on several of its other moons, including Enceladus. The two moons are both fascinating but, right now, we're going to focus on Enceladus, so let's dive in and have a closer look.

Figure 9.2: Size comparisons between Saturn (one of Rundle Mall's silver balls), Enceladus (pea, shown here on a dinner plate), Earth (basketball) and our Moon (tennis ball).

Enceladus is one of the major inner moons of Saturn. It is the sixth-largest (both in terms of size and mass) and, as mentioned earlier, is much smaller than our own Moon. Just how much smaller?

It's approximately one-seventh the size of our own Moon, but it can be hard to get our heads around those sorts of numbers, so let's try and put that into perspective. Imagine that Earth was a basketball. At this scale, our own Moon would be roughly the size of a tennis ball, Saturn would be about 2 metres in diameter (about the size of one of the Rundle Mall balls—Hi, Adelaide!) and Enceladus would be just a tiny little pea. Figure 9.2 shows a photo of what that looks like (for those outside South Australia). Isn't it fascinating that, even though Saturn is so much larger than Earth, our Moon is bigger than Enceladus? As we mentioned in Chapter 5, our Moon is the biggest moon in the Solar System *relative* to the planet it orbits. So, while there are bigger moons orbiting Jupiter, they're much smaller compared to the size of Jupiter than the Moon is to Earth.

WHAT'S HIDDEN UNDER THE ICE?

With Enceladus, we've got a decent-sized moon orbiting Saturn, so let's start digging into this a bit deeper. *Literally*. Let's talk about the inner structure and composition of Enceladus.

What is Enceladus made from? As we mentioned earlier in this chapter, Enceladus is cold, icy and pretty far away. *Voyager 1* and *Voyager 2* performed flybys of Enceladus; during the latter, we obtained imagery of its surface revealing various geological features, such as craters, plains (called *planitiae*) and ridges (called *dorsa*). However, once *Cassini* was in orbit about Saturn, we really started to get a better understanding of Enceladus. While *Voyager 2* observed various surface features (and indeed, so did *Cassini* in greater detail), something dominated the surface of the planet: smooth, clean ice. Check it out in Figure 9.3.

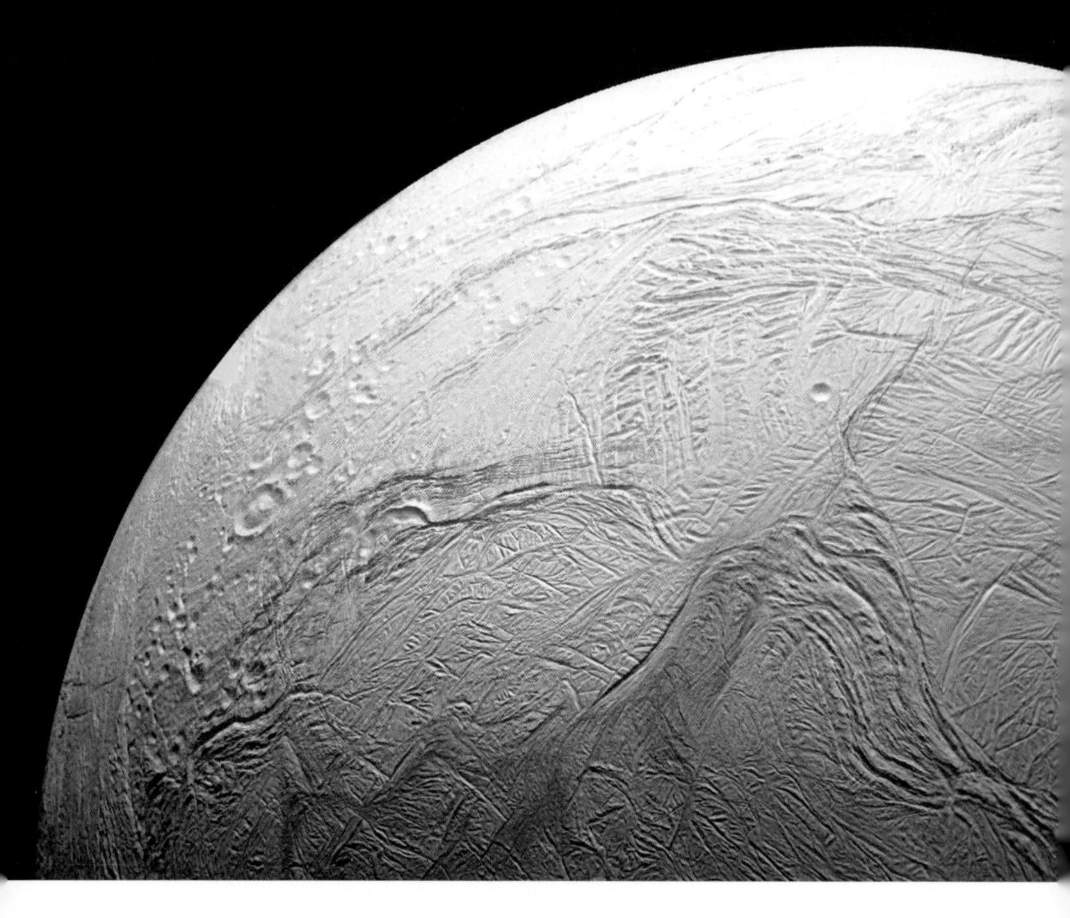

Part of observational astrophysics involves taking observations like these and recognising the greater implications. Here, we have a moon that most likely experienced similar levels of impacts as we see in the cratering on bodies like the Moon or Mercury, and yet on Enceladus, we have a much smoother surface. There aren't nearly as many craters. So what happened? The scientific explanation is that recently (geologically recently, meaning, in the last 100 million years), Enceladus experienced a sort of *water volcanism*. That's right. *Water volcanoes!* Or *cryovolcanoes*, if we want to be scientific. Just as molten lava on Earth will cool and solidify into rock, as this water bursts out of the surface of Enceladus, it spills onto the surface and cools into solid ice. This 'resurfacing' or 'refreshing' of the outer ice of Enceladus is what ultimately yielded its smooth, clean, icy surface. It could fill in and smooth out craters from previous impacts. This discovery helped to explain another question scientists had grappled with regarding

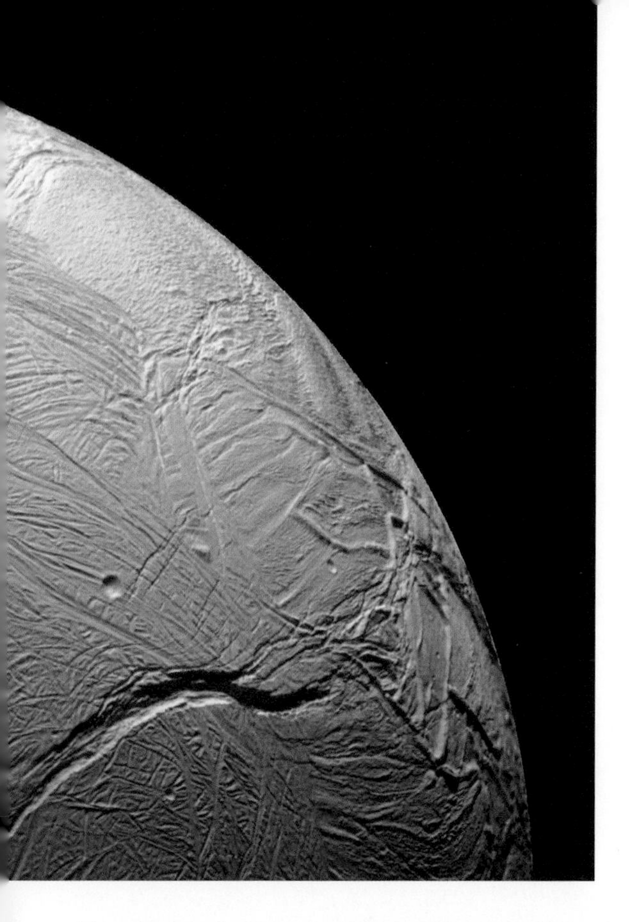

Figure 9.3: Enceladus as captured by *Cassini*. Images like this are often many images stitched together (known as composites). The images are taken in different wavelengths and this information is then used to colour the image according to what we think it would look like.

Enceladus for years: Why is Enceladus so bright? The answer comes from this same phenomenon. The smooth, icy surface is highly reflective (a property we refer to as having a high *albedo*) and, because of this, Enceladus shines extremely brightly as it reflects light from the Sun.

What's under the icy surface of Enceladus? Was this water volcanism analogous to what happens here on Earth with rock? Is Enceladus just a ball of ice that once had a liquid water core in the geologically recent past? Ultimately, *Cassini* helped us answer these questions, along with planetary formation models. We discussed planetary formation in Chapter 4 and, while we won't re-open that complicated can of worms, we will dredge up the critical part that applies here; in general, objects formed far out in the Solar System—at the distances of Saturn, Uranus and Neptune—will form with a rocky core that then grows quickly with water ice. We can then

understand the composition, or proportion, of the rock and water by assessing the object's mass and size (rock is denser than water and so, for two objects of equal mass, the rockier one will be smaller than the more watery one). After conducting such an assessment, we determined that this is true for Enceladus: the icy surface envelopes a rocky core.

So, there's solid ice and a rocky core. However, the question we're dying to answer is: Is Enceladus still geologically active, or was it active only in the past? Underneath the ice, is there still liquid water on Enceladus, or has it all cooled so much that the water volcanism is no longer happening? Well, our trusty friend *Cassini* helped us solve this

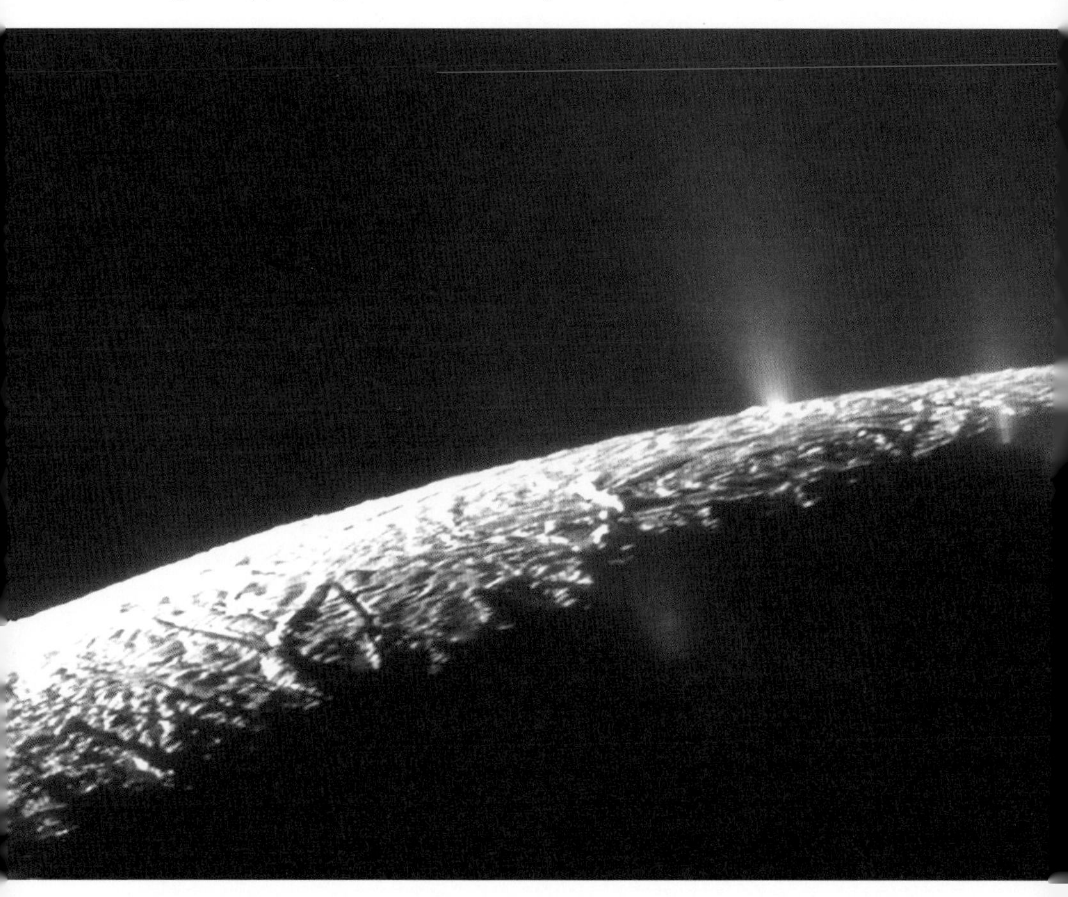

Figure 9.4: The 'water volcanoes' of Enceladus. Scientifically, they are referred to as *cryovolcanoes*.

problem, too. *Cassini* detected enormous water plumes bursting out into space, some extending as far as 500 km from Enceladus at speeds of up to 3.4 km per second. These are huge water eruptions—take a look at them in Figure 9.4—and detecting that these are happening even today was (and still is) a major discovery, because it paints the picture that Enceladus is still geologically active. Because Enceladus is so geologically active, the accepted internal structure is that, sandwiched between the outer icy shell and interior rocky core is an abundant, liquid water ocean. Liquid water exists, today, on Enceladus, just below its icy shell.

ENCELADUS'S COSMIC DANCE

We also know that Enceladus is cold, so how can there be liquid water? Surely, it would be frozen solid this far out from the Sun? Enceladus is about –200°C. It's mind-numbingly cold. The only way for liquid water to exist this far out is for there to be another source of heat besides the Sun; something else is responsible for generating the energy needed for water to exist as a liquid. It can't be another star; Saturn is pretty far from the Sun (a little over 1 light-hour—that's the distance light travels in one hour, and light's pretty fast. For some context, it takes light from the Sun 8 minutes to reach Earth) but the next closest star system is more than 4 *light-years away.* (As you may've gathered, a light-year is the distance light travels in one year. Based on the number of hours in a year, we can calculate that the next closest star is more than 35,000 times further from Saturn than the Sun!) No, the source of heat is somewhere much closer to Enceladus. The heat source is actually inside Enceladus itself. Referred to as *tidal heating*, it occurs on several moons in the Solar System. Tidal heating is heating generated through tidal friction. If we can understand tidal friction, we'll understand what causes tidal heating, so let's break down the two terms.

Remembering back to Chapter 5, we discussed what tidal forces were, and how they relate to the tides here on Earth. As a quick refresher, the rising and falling of the tides happen because of the Moon's gravity, specifically, that gravity decreases the further we are from an object.

This same principle can occur on other bodies in space, such as moons like Enceladus, but when multiple objects get packed tightly within close proximity to one another, or really close to a massive body such as Saturn, some even wilder stuff starts going on. In the case of Enceladus, it's in a sort of dance with another moon, named Dione. In astrophysics, this dance is called an *orbital resonance* and it just means they like to be in sync when they orbit; every time Dione orbits Saturn once, Enceladus orbits Saturn twice. This dance results in the same tidal bulging on Enceladus as I described earlier with the oceans of Earth. The difference, however, is that the rocky structure of Enceladus is experiencing 'tides' rather than a freely flowing fluid like an ocean. On Enceladus, the tidal bulging causes stretching, pulling and grinding deep within its core. Now we've come to the 'friction' part of tidal friction. Just like when you rub your hands together, grinding friction creates heat. *Lots of heat.* And this heat, generated by *friction*, because of the *tidal* stretching and bulging of the core of Enceladus, is the same tidal heating that we introduced as the source of energy needed to maintain a liquid water ocean. And it is precisely this subsurface liquid ocean that we're interested in. By understanding the water jets (or water volcanoes) exploding from the surface of Enceladus into space, we can begin to uncover what's lurking within this subsurface ocean *without actually looking in the ocean itself.*

FLYING THROUGH AN ERUPTION

As mentioned earlier, *Cassini* detected (and took pictures of) the water volcanoes of Enceladus, but that wasn't the only thing *Cassini* did. In a serendipitous set of circumstances, *Cassini* actually flew through these plumes of water stretching out from the surface. This meant *Cassini* could utilise some of its instruments (specifically the ion and neutral mass spectrometer, and the cosmic dust analyser) to measure the composition of atoms, molecules and charged particles in the water jet, which would indicate the composition of the subsurface ocean.

So, what exactly was detected? Obviously, water vapour dominated the detection—it's a jet of water, after all—but other molecules were present as well, such as nitrogen and carbon dioxide. However, the most exciting and profound detections were those of low-mass organic compounds that are nitrogen-bearing, oxygen-bearing and aromatic. Now the term 'organic' isn't the exciting word here. Often, 'organic' is sensationalised—particularly in the media—to mean *organism*, or *life*, but when referring to chemistry, organic has to do with the chemical structure of a compound. Organic means it is carbon-based with carbon–hydrogen bonds; while it's still an exciting discovery, we have already detected organic compounds elsewhere in the Solar System and space. No, the exciting thing here is that these particular compounds look to be progenitors to *biologically relevant organic compounds*. Now the word 'biological' is not being sensationalised here. Biological simply means pertaining to biology or living organisms.

This is the most impactful and exciting part of this whole discovery. It looks like the ocean of Enceladus is an environment that has all the ingredients for life.

THE ENCELADEAN NIGHTS

Here's a fun little fact. The features of Enceladus are all named after people and places from the famous compilation of Middle Eastern folk tales, *One Thousand and One Nights*, translated into English as *The Arabian Nights*. Some craters named after people include the *Aladdin* and *Sindbad* craters, while some sulci (long grooves along Enceladus's surface) are named after places, including Baghdad Sulcus, Cairo Sulcus and Damascus Sulcus.

THE INGREDIENTS FOR LIFE

What does that really mean though: 'An environment that has all the ingredients for life'? We know that—loosely speaking—Earth is an environment that has all the ingredients for life. We're here. But more specifically, what does it mean? My PhD is in astrophysics not biology, but there is a small overlap between the two (astrobiology and astrochemistry); I learned some of the fundamentals, specifically: the requirements for life. As we discussed in Chapter 6, life can be challenging to define, but we have determined the requirements (at least for life as we know it) a little more clearly—outlined in Chapter 7. In essence, we need two things. The first two are:

1. liquid water
2. an energy source.

In addition to this, to kickstart life (as we highlighted when we discussed the Primordial Soup hypothesis in Chapter 6) we need the third ingredient:

3. biogenic elements, which are elements we find consistently in living organisms (notably carbon, oxygen and nitrogen).

For most life on Earth, this stuff is easy. Liquid water is all around us in many forms— oceans, seas, rivers, rain and water vapour in the air. Energy source? The massive nuclear reactor in the sky takes care of our energy needs (at least at the highest level), which is often converted to other forms of energy by plants through photosynthesis. This creates the means by which we usually access this energy: eating food—vegetables or animals if you're an omnivore. Again, biogenic elements are quite abundant. Nitrogen and oxygen dominate our atmosphere, and carbon (as well as plenty of heavier elements, for that matter) make up our Earth.

What about when these things aren't as obvious? On Earth, such an environment is on the ocean floor, deep below the ocean's surface. It's so deep that the energy from the Sun doesn't penetrate. Deep underwater on the ocean floor, the environment appears to be missing one of the key ingredients for life: energy. Furthermore, it may also be deficient in the right biogenic elements. Enter black smokers (which we touched on in Chapter 6). These are hydrothermal vents on the ocean floor, which get their name from the black smoke that billows through the vents and, fortunately, they provide these missing ingredients.

A quick recap on black smokers: The Earth's core is home to all sorts of violent geological activity, which makes it immensely hot and highly pressurised. This energy can often make it to the surface in the form of volcanoes (above water), or hydrothermal vents (underwater). In the case of hydrothermal vents, the energy spews forth as heat, bubbling up through the vent and bringing with it minerals from deep within Earth's core. And I bet you can guess what sort of elements are in these minerals? The biogenic kind! Thus, deep on the ocean floor, where we are missing the usual trifecta of ingredients for life, hydrothermal vents have it covered by pumping energy and biogenic elements into the water.

As we mentioned earlier, wouldn't you know it, in this environment that has all the ingredients for life, *we find life!* The most incredible thing about this life, however, is that it's strikingly different from what we see near Earth's surface, where the energy source is the Sun. Here, we find unique life forms, such as *tubeworms*, that don't harness energy from the Sun, but energy from the planet. This environment—which has a fundamentally different energy source to environments near Earth's surface—manages to support life.

It is this flavour of environment that we believe may exist on Enceladus. Because the Sun's energy has dissipated far too much by the time it reaches Enceladus, internal tidal heating must be responsible for the liquid ocean. The detection of *biologically relevant organic compounds* in the water jets of Enceladus seems to be a strong indicator that there may be some sort of geological activity there, too, like black smokers. And around black smokers on Earth, we find life, so, well—you see where I'm going with this?

That's what is mind-blowing about Enceladus. Utilising some brilliant ingenuity, we launched a spacecraft from Earth, slingshotted it around several planets to arrive at its destination—orbiting Saturn. While there, a little icy moon with an internal heating mechanism and subsurface ocean had a water eruption, spewing forth an enormous jet of water that our little spacecraft passed through. After measuring the chemical composition, it found chemistry similar to that of the black smokers here on Earth, the very black smokers around which we find life.

There's been a lot to unpack but it's so exciting, and it highlights that here, in our own Solar System, we've found somewhere that might support life.

EUROPA

What other moons are of interest? Well, another one plays a similar game to Enceladus, which is one of Jupiter's moons, Europa. Unlike Enceladus, Europa is similar in size to our Moon (about 90% the size of our Moon) but Europa is still much smaller, relatively, than its host planet, Jupiter. The interesting thing about Europa is that it's in some big company. Jupiter has 53 named moons and more than two dozen more waiting to be named, a total of 79 known moons (as of 2022).

Four of Jupiter's moons make up an interesting subset, however;

they are the *Galilean moons*, aptly named after Galileo Galilei, who discovered them at the start of the 17th century with a magnifying telescope. The Galilean moons are special because they're the innermost moons that orbit Jupiter and they're big. Along with the Moon and Titan (a moon of Saturn—the target of *Cassini-Huygens*), the four Galilean moons round out the six largest moons in the Solar System. The largest, Ganymede, is so big, it's larger than Mercury, one of the planets in our Solar System! But I digress; let's get back to our main focus, Europa.

Europa is big, it has some even larger neighbours and, like Enceladus, it's cold. Not quite as cold—Jupiter is only about five times as far from the Sun compared to Saturn (which is about ten times as far)—but the surface temperature of Europa still sits at a brisk –170°C. This means Europa also appears as an icy ball, and faces the same problems as Enceladus: the Sun's energy is not sufficient to ensure liquid water can exist. However, there's a growing body of research, coming from a string of scientific observations, that seems to indicate that there is indeed liquid water on this icy moon.

MAGNETIC CLUES

The first evidence of liquid water came in the form of changes in Jupiter's magnetic field. Just as we have a magnetic field here on Earth, so does Jupiter. Over a period of nearly a decade around the turn of the millennium, we observed changes in Jupiter's magnetic field near Europa. If you recall from Chapter 2, we discussed how electricity and magnetism are inextricably linked; they're two sides of the same coin. As such, changes to the field of one (an electric or magnetic field) impacts the other.

To explain the weird changes in Jupiter's magnetic field, scientists suggested that, beneath the icy shell of Europa, the presence of a fluid that could conduct electricity (such as a salty ocean) was causing these magnetic fluctuations. This was an excellent hypothesis that fitted what we observed, but it was still an early hypothesis that was inferring the presence of a subsurface ocean. So, was there something more obvious we could detect? Our answer came in 2013, when the Hubble Space Telescope observed the chemical constituents of water (hydrogen and oxygen) in what looked like plumes or jets in Europa's atmosphere. In 2016, these structures were then photographed as silhouettes with the canvas of Jupiter in the background. These detections are strong evidence that jets of some chemical substance composed of hydrogen and oxygen are surging upwards from the surface of Europa – and we've seen these before, right? Cryovolcanoes!

The final confirmation came in 2019, when scientists used the Keck telescope in Hawaii, USA, to detect water on Europa spectroscopically. Remember our dancing eggs analogy in Chapter 7? We use spectroscopy to detect chemicals, such as water, by analysing light. Different chemicals use different amounts of energy to make their eggs dance in different ways, and these different amounts of energy will correspond with specific colours. Depending on which colours of light are missing, we can determine which chemicals we are looking at. Essentially, this team of scientists discovered that the chemicals (eggs from our analogy in Chapter 7) exhibited the absorption lines of water (they were doing the water dance), so they can say with confidence that they detected water.

ORBITAL DANCING TRIO

How does liquid water exist on a body so far out from the Sun? Well, not only is the structure of Europa eerily similar to that of Enceladus, it has a similar mechanism generating heat. The three innermost

DON'T FRIGHTEN THE TITAN

Cassini-Huygens was destined for Saturn, and the *Huygens* probe was intended to descend through the atmosphere of Titan and land on its surface. The probe achieved its goal in 2005 and the imagery the *Huygens* probe obtained from Titan's surface is incredible. But I think the *Cassini* space probe obtained some images that are even more profound.

At this distance, it's too cold for liquid water to exist on the surface of planets, so while Titan is one of a small group of rocky bodies with an atmosphere, it's too far from the Sun to have liquid water on its surface (subsurface oceans are a different story, as we've discussed with Enceladus and Europa). It's not too cold for *all* liquids to exist, however. In 2007, scientists were tremendously excited when they confirmed a long-held hypothesis about Titan—there were lakes on its surface! Now, these were liquid *methane* lakes, rather than liquid water lakes, but they're big lakes, as big as the Great Lakes of North America.

This was a profound discovery: the first liquid lakes discovered off Earth. A different liquid than what we have in abundance here, but lakes, nonetheless. This discovery raises the question: Is it water, or just a liquid, that is critical to life? In an earlier Fun Bubble, we discussed how we consider all life to be carbon-based, even though there may be other forms of life using other elements as building blocks. Water may be similar; perhaps there are wildly exotic forms of life that are not only swimming about in these methane lakes but that are methane-based in the same way that life on Earth is water-based. In Chapter 1, we highlighted what makes water so unique and special, and so maybe the notion of life built around other liquids is just silly. There's no way life could exist in the lakes of Titan. *Or could it?*

moons of Jupiter—Io, Europa and Ganymede—are locked in what we call a Laplace resonance, a type of orbital resonance similar to what Enceladus is in with Dione. In this scenario, however, all three moons are synchronised so that, for every four orbits Io completes of Jupiter, Europa completes two, and Ganymede completes one. This is a 4:2:1 resonance; it is incredibly stable and ensures that the inner Galilean moons remain in a regular, stable orbit around Jupiter. These moons, and Europa in particular, experience the same tidal forces that Enceladus experiences. The bulging, flexing and stretching that is the gravitational tug-of-war among the Galilean moons and Jupiter results in the interior of Europa to experience flexing, grinding and friction, which causes the inner core to be heated. This heat leads to the liquid subsurface ocean on Europa; it provides enough energy for liquid water to exist at this distance from the Sun.

As we've discussed, there is strong evidence that Europa has a vast, subsurface ocean beneath its icy shell. Just as we analysed the jets from Enceladus to identify whether they have the right chemistry for life, we are trying to gather similar sorts of evidence from Europa's jets (which we will dig into in greater detail in Chapter 13). Just think about it though! We now know that two moons—Europa and Enceladus—have subsurface oceans and blast jets of these oceans into space, and (in Enceladus's case) we've analysed the jets to identify that the ocean is chemically rich. So, Mars is not the only rocky body harbouring liquid water. There are some pretty funky moons orbiting the big gas giants that are remarkably promising for having all the right ingredients for life to exist.

This is an incredible idea to think about. Even within the confines of our own Solar System, there are places where liquid water exists and which might have the right ingredients for life. If life happened on Earth, why not in one of these other environments? The effort to discover if life has existed, or continues to exist, in these environments is a huge undertaking, but it's integral to our understanding of life and its origins. In Part 1, we focused on Earth and, so far in Part 2, we've been looking for locations within our own Solar System. So the logical next step is for us to cast our net even wider. Let's look beyond the Solar System at planets orbiting stars other than our Sun.

CHAPTER 10

EXOPLANETS

Up until now in this book, we've discussed things in a Copernican fashion. What does that mean? Here's a turbo-charged history lesson. Humanity's original understanding of the heavens was geocentric. Earth was—or more accurately *we were*—the centre of the entire Universe (which, at that time, was the Solar System with a lovely backdrop of stars and nothingness) and everything revolved around us. In the 16th century, Nicolaus Copernicus proposed a radically different hypothesis, suggesting that the Sun is actually at the centre of the Solar System, and we orbit around it.

I won't go into all the subtle nuances and intricacies of this revolutionary idea here, but the analogy I want to borrow is that this book so far has been Copernican: we've looked at objects that are in orbit around the Sun (the Earth and the Moon, Mars and some of the more interesting moons orbiting Jupiter and Saturn). More recently in astronomy, we have uncovered something that we were quite certain existed but had yet to detect ourselves. That 'something' is, of course, planets outside our Solar System, so-called *extrasolar planets* or *exoplanets*. Considering the number of potential locations where liquid water exists just in our own Solar System, you can imagine the sheer enormity of the discovery of planets in completely different star systems and potential locations to search!

MAIN SEQUENCE STARS

Let's have a quick history lesson about exoplanets: how did they go from something we hypothesised existed, to finding evidence to confirm our hypothesis, all the way through to finding a slew of exotic, yet Earth-like, worlds? As you might imagine, there's a lot of detail and nuance to this story, so I'm going to concentrate on a subset of star that we're most interested in. We're going to focus on those planets that orbit one particular type of star: a main sequence star. That then begs the question: what *is* a main sequence star?

Main sequence stars are what you might consider your run-of-the-mill star. Not that all stars are the same, but they're similar in the way you might consider someone you meet as a main-sequence human. You might meet some humans who are pushing a pram (they're main-sequence humans); they might have smaller, little humans in their pram (they too are main-sequence humans); and they might be walking down the road to a cute little café to meet grandma and grandpa for a slice of sticky date pudding (they too—the grandparents, not the pudding—are also main-sequence humans). The point is that all these humans are on the main-sequence, and the main-sequence is a *sort of* timeline for stars. I say 'sort of' because it's not entirely true, but it's useful for our description here.

Stars exist on this main-sequence timeline for a period of time that's almost entirely based on how massive the star is. Small, low-mass stars will burn very slowly and stay on the main-sequence for a *really* long time. We're talking longer than the current age of the Universe! This is because a star that weighs less also burns slower;

there's less gravity crushing everything together deep inside the core of the star, which means that nuclear fusion—the reaction that causes the explosion and release of huge amounts of energy (we discussed this briefly in Chapter 1)—takes places at a much slower rate. The rate is slower because fusion reactions occur randomly when chemicals encounter one another. Let me use an analogy to explain just why things being crushed less densely together means fewer reactions.

Imagine that I have ten basketballs bouncing around in a huge basketball stadium. They're quite unlikely to hit each other, whereas if I have those same ten basketballs bouncing around in, say, a tiny bedroom, they will likely have loads of collisions with one another. Lighter stars only have enough mass to squash all their fuel into the huge stadium because there's not enough mass for gravity to squash it into a smaller space, so their reactions happen less frequently. In contrast, the largest, brightest stars burn the quickest. From how we described the slow rate of burning, you should be able to hedge a guess why that might be the case. That's right, the larger stars have enough mass and hence gravity to crush all the fuel into a smaller, tighter space: the basketballs are in the bedroom. The more mass there is, the more gravity there is, crushing everything together deep inside the star's core, causing nuclear fusion to take place far more rapidly, leading to huge amounts of energy being released (some in the form of light, hence the increased brightness).

Most stars are main sequence stars. When the chemical fuel that sustains these explosive reactions inside the stars starts to run out, different things start to happen, and that's where our analogy of the main-sequence as a timeline falls apart a bit. It's beyond the scope of this book but, essentially, when the basic fuel runs out, you get all sorts of exotic forms of stars starting to form:

- **Red giants** These tremendously bright stars are enormous in size. (For comparison, if the Sun was a red giant, it would stretch beyond the orbit of Venus.)
- **White dwarfs** These stars have exhausted their fuel and no longer have explosions taking place inside their core, but they are still quite hot, so they are slowly dissipating light as they cool.
- **Neutron stars and blackholes** When huge stars collapse in on themselves into tiny volumes, these two bizarrely exotic stars form, creating some of the most perplexing and strange environments we know of.

While this is far from an exhaustive list of weird and wonderful stars (for example, *pulsars*, *magnetars* and *supergiants*, oh my!), I want to focus on our main-sequence timeline analogy. Most stars exist on this main-sequence; they go through their burning at different rates, depending on their mass and chemical composition, then they eventually leave the main-sequence to become something different, depending on what was going on with their mass and chemistry. We are most interested in those on the main-sequence; first, because they're the most abundant—we may as well look at the most common and abundant stars—and second, because (as you might have guessed) the Sun is a main sequence star. If we're going to look for planets like Earth, and places where life could exist, we need to start with the basics: those places that most closely resemble the only place we know of where life exists: our Earth orbiting the Sun.

In the unending vastness of space, we gaze up at the beauty of innumerable stars that light up the night sky. Each of these scintillating gems woven into the black canvas of space represents a star, and these are just the ones we can see with our naked eye. Our eyes also adjust to ambient light—you've probably noticed that when you get out of the city, the night sky looks so much more impressive. There aren't more stars, they're just more easily visible because

ON THE SHOULDERS OF GIANTS

When we say the environments around neutron stars and blackholes are bizarre, what we're referring to is that the gravitational pull around black holes, right up close to them, causes our understanding of physics to break down. Okay, I'm being a little sensational here for effect, but the point remains. There are certain things blackholes do that just don't make sense or fit with our current understanding of physics. This is a really important, and humbling, point to remember in science. Physics and science are about learning to understand and describe the world around us, but there are still things that exist today that just don't work in terms of our current understanding and explanations.

In school, we could check to see if we got the answer right and, if we didn't, we could go back through our calculations and theory to understand what we missed. But we're figuring this all out as we go, and that's the wonder and beauty of research. You're trying to learn, understand and describe the world around you in a way that no one has done before. You're adding to the body of knowledge that, until you add to it, *straight up does not yet exist!*

Now, back to things in our world that our current understanding doesn't quite describe correctly: blackholes. That our current models can't describe some of a blockhole's behaviour isn't a failure. It is progress! It shows there's more to find out. When Albert Einstein proposed his General Theory of Relativity, for example, it didn't break Isaac Newton's proposed description of gravity, or suggest what Newton had put forward was incorrect. What it did was introduce a way to further complete the picture. It found the gaps that had thus far been unexplainable, and hypothesised a new, more complete way to make it work.

our eyes have adjusted to allow more light in. Of course, we've used human ingenuity to get around the limitations of the human eye and look into space with advanced technology (we'll get to that in much greater detail in Part 3), but let's get back to the night sky and the abundance of main sequence stars in it.

To fill in the blanks; to give a more complete picture of our understanding of the Universe: this is the beauty of science. Who knows what, or when, the next hypothesis will be put forward to keep filling in the blanks that we keep finding. I promise you, we are very far away from 'finishing science'. We're not on the last level of Super Mario, about to save Princess Peach from the final castle. There's still so much to be explained, and so much to be uncovered, that could be just one 'Hey, that looks a bit odd,' observation away. Each generation of scientists builds on, adds to, further refines and ultimately pushes forward our understanding of the world around us. As Isaac Newton once eloquently put it: 'If I have seen further it is by standing on the shoulders of Giants.'

UH OH, SPAGHETTI-O!

The environments around neutron stars and blackholes are bizarre, but one spectacular aspect is referred to as *spaghettification* (one of my all-time favourite words in astrophysics!). At several points in Part 1, we highlighted that the strength of gravity decreases with distance. For someone standing on a planet, this effect is negligible; the strength of the planet's gravity pulling on your feet will basically be equal to the strength pulling on your head (not exactly equal, but the difference is so inconsequential that it's considered identical).

Near a blackhole, however, is a different story. Because a blackhole is so tiny (it's actually a point in space—a *singularity*—but this is more than we can cover in a Fun Bubble), there is a *huge* difference between the amount of gravity your head and feet experience as you get close. That difference becomes significant in the scheme of the overall distance from the point in space where the blackhole is pulling from.

To get our heads around this concept, let's put some numbers together. On a planet with a radius of 6371 km, or 6,371,000 metres (Earth), you can only stand on the surface, and so the difference in gravity between your feet and your head due to your height of 2 metres (if you're a big unit!) is negligible, because your feet are 6,371,000 metres from the centre of Earth, and your head is 6,371,002 metres (the radius of Earth adding your height). The difference in gravitational pull is roughly 0.00003%. That's the same scale as about a litre in an Olympic-size swimming pool. If you scoop out a litre of water from the pool, it basically has no effect.

As you approach a blackhole, however, the 'planet' or singularity is so small your 2-metre height can be *quite* significant. Your feet could be 2 metres from the blackhole and your head 4 metres. Your head is twice as far away! Because you can get so much closer to the singularity of a blackhole, you can experience a gravitational pull on your legs that is 2 times, 10 times, even 100 times stronger than on the top of your body—what this does is stretch you out. Your lower body is being pulled stronger and harder towards the blackhole than your upper body. It's like if you held some dough in your left hand and pulled it with your right. The dough is stretched. A blackhole stretches you out in much the same way, only it doesn't stop. It stretches you so much that you're pulled into a long, thin piece of spaghetti; hence, *spaghettification*. Doesn't sound pleasant at all!

WHERE ARE THE EXOPLANETS?

How do exoplanets fit into all this main sequence stars malarkey? For quite some time, we have hypothesised that planets could exist outside our Solar System. In the same way that early European explorers hypothesised the existence of the 'New World' and 'Terra Australis', scientists and humanity similarly hypothesised the existence of planets orbiting stars other than our own Sun.

Curiosity is a quintessential property of the human condition, and before we have the ability to satisfy our curiosity in practical ways, we often resort to logic: 'If *this*, then surely, logically, *that*'. If we exist on this land, then surely there is another land we don't know about, where others surely must also exist, pondering the exact same questions we are. This curiosity, this thirst to understand our place—not just in the world or the Universe, but among other living things—encapsulates humanity's unrelenting pursuit of knowledge, and our unyielding development of greater and more sophisticated technology. Technology to help us uncover the answer to who we are, are there others, and how we fit into all of this.

While we had an inkling there must be planets orbiting other stars, it wasn't until 1995 that our curiosity itch was finally scratched. There's a star located approximately 50 light-years away called 51 Pegasi. (We mentioned light-years last chapter—a measure of distance equal to how far light can travel in one year. For some context, *Voyager 1*, the spacecraft that is the furthest from Earth, launched more then 43 years ago and travelling at approximately 17 km *every second* is *almost* 1 light-day away.)

It turns out, 51 Pegasi is not only a main sequence star, it's a Sun-like main sequence star (we won't go into all of the details of the main sequence stars but there are sub-classes, or spectral types, of main sequence stars, and 51 Pegasi is the same spectral type as the Sun: a G type).

Now, 51 Pegasi (shortened to 51 Peg) had been observed for quite some time via telescope in France, specifically, using an instrument called a spectrograph, which looks at different wavelengths, or colours, of light being emitted from an object. This is important because what you notice when observing the wavelengths, or colours, of different parts of light from a star is that, well, *they move*. They don't just move randomly or erratically either, they move in a controlled, cyclical and repetitive fashion. The movement of these lines of colour, or what we refer to as *spectral features*, is the basis of the first of the two main techniques that have successfully detected exoplanets: the *radial velocity method*.

What the French scientists, Michel Mayor and Didier Queloz, had successfully uncovered was the first planet found outside the Solar System orbiting a main sequence star. Arguably even more mind-blowing, was that the technique they used was so elegantly simple, yet so widely applicable to other stars, that they opened up a new era of astrophysics (which I myself am most definitely guilty of being drawn into). This has gripped the curiosity of thousands of scientists, and exploded in both the areas of public interest and academic research within the astrophysics community and beyond.

THE RADIAL VELOCITY METHOD

We need to sit down and unpack what's going on with the radial velocity method. What is it and how does the to-ing and fro-ing of these spectral lines somehow indicate the presence of a planet? To understand what's going on here, we need to introduce a few fundamental physics building blocks. Each one will come together to ultimately tell the story of how Mayor and Queloz discovered the planet 51 Peg b,[4] for which they received a Nobel Prize in 2019. (Science is a powerful, rigorous and profound beast, but things take time; it can often be several decades between scientific discovery and a formal honour like a Nobel Prize.)

The first thing to introduce (or re-introduce, really) is our dancing eggs from Chapter 7. We've discussed how these dances can be used to indicate which chemicals are present, because of which colours are missing in the spectrum of light. The missing colours are like the signature, or fingerprint, of different chemicals. Water has a particular set of dances, and the energy and colours corresponding to each of these dances will be missing from the spectrum in a unique way. So too does sodium, iron, neon—whatever the element or molecule, there is an associated 'fingerprint' corresponding to that chemical's dance. We also know from Chapter 7 that light can be split into a rainbow, or spectrum. Because light can be split into a rainbow, it's easy for us to see where the missing light in the spectrum is: we call these the *spectral lines*.

These lines correspond to our analogous eggs' dance moves. This is the first thing to note when we absorb the light from a distant star; even though it may appear to be essentially white, at a detailed level, there are missing colours and missing spectral lines. Figure 10.1 shows a comparison between a perfect, continuous spectrum (top) and the absorption spectrum for different elements (hydrogen and oxygen), showing how each set of lines are unique to a specific chemical.

I'm just going to park the idea of spectral lines for a moment. We'll come back to them shortly but, first, we need to introduce another idea to give some context so it all makes sense: the *Doppler effect*. The Doppler effect is something you would have almost assuredly already experienced in your life.

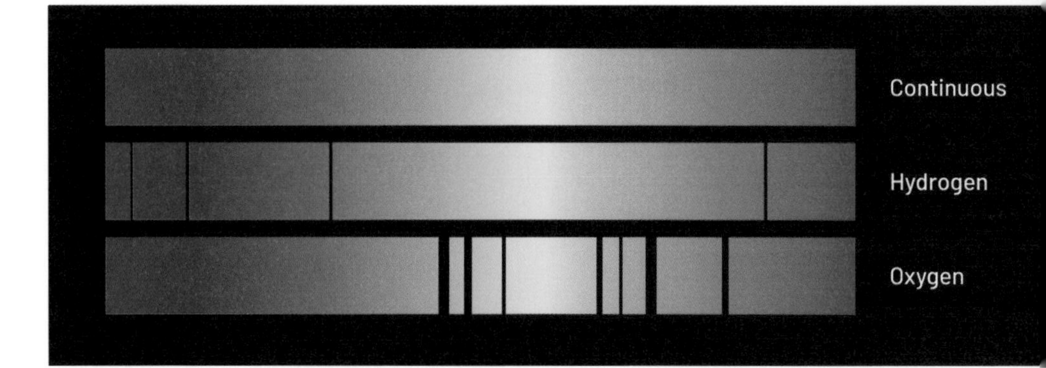

Figure 10.1: A comparison between several spectra to show how different chemicals have different spectral features. When white light is split it will show as the continuous spectrum like a rainbow (top), whereas if an element like hydrogen (middle spectrum) or oxygen (bottom spectrum) absorbs white light, the light that has transmitted through will be missing certain colours when it is split, and these show up as spectral features or lines.

Have you ever noticed that, when a car (especially if that car has a siren like a police car or ambulance) drives past you, the sound of the engine (or siren) has a higher pitch as it approaches you, then it starts to have a lower pitch as it passes by and drives away from you? That is the Doppler effect. It is the phenomenon whereby waves (sound waves, in this scenario) change their frequency based on the motion of the object generating those waves. This is because the waves in front of the object get 'squashed' as the object moves into them ('them' being the sound waves) as the object approaches you, while the waves behind the object get 'stretched' as the object moves away from them ('them' again being the sound waves) as the object leaves you.

Figure 10.2 does a great job of demonstrating this phenomenon, using the waves of water from when a duck swims. Our adorable little duck is swimming on a lake and, for clarity, we've overlaid the wavelengths (that is, the distance between each wave) of the ripples in front and behind the duck. We can see that the ripples (or waves) in the direction that our ducky friend is swimming in are being squashed and compressed (shown in blue), while the ripples (or waves) behind it are being stretched (shown in red).

Figure 10.2: You can see the Doppler effect when a duck swims through water. The waves (or ripples) formed by the duck are squashed in the direction the duck is travelling towards (the blue lines), while the waves (or ripples) are stretched in the direction the duck is travelling from (the red lines).

So, what do adorable ducks have to do with finding planets? Well, it all boils down to the Doppler effect, which occurs in all forms of waves. We've talked about sound with cars and sirens, we've talked about water with our swimming duck, but what about light waves? This is where it all starts coming together! Just like those other waves, light waves also experience the Doppler effect. When something emitting light moves towards you, the light gets squashed and the frequency increases. Likewise, when something emitting light moves away from you, the light gets stretched and the frequency decreases.

The ripples around the duck are coloured intentionally, because of what it means when the frequency of light changes. When light is squashed—because the object emitting the light is moving towards you—the frequency of light increases, which corresponds to light being slightly bluer. In contrast, when light is stretched—because the object emitting the light is moving away from you—the frequency of light decreases, which corresponds to the light being slightly redder. These two phenomena are so important in astrophysics, they have their own terms associated with them: blue-shift and red-shift. Red-shift in particular is a common turn of phrase as it means that things are moving *away* from us.

Have you ever heard that the Universe is expanding? It's the red-shift of distant galaxies that tell us this is the case! Instead of appearing more or less white, they are slightly red. If everything we look at has its light slightly redder than it should be, it means that, relative to us, those galaxies are *moving away*. The discovery that the Universe is expanding because everywhere we look objects are slightly red-shifted was huge back in the early 1900s and still has profound implications today, but we'll save that for another book. The salient point here is that, if the colour of an object has been red-shifted, it means it's moving away from us.

We're almost there now. We've got all the pieces of the puzzle: a method to detect exoplanets called the *radial velocity* method; the signature, or fingerprint, of a star that shows unique spectral lines; the Doppler effect that shows how waves change frequency, depending on the motion of the source of the waves; and the application of the Doppler effect which, with respect to light, results in a reddening or blueing of light.

There's just one more thing we need to discuss: What it means for a planet to orbit a star. Contrary to what you might believe, when a planet orbits a star, it's not like the star is stationary and it is just the planet that is moving. It is a little more complicated than that. Let's just consider a star with a single planet for a moment to describe what is actually happening. Rather than the planet orbiting around the star, it is actually the two bodies (that hold onto each other by their gravitational attraction) orbiting around their common centre of mass, which is sort of like the middle of two masses. If two masses are equal, the centre of mass will be smack bang in between them. If one mass is bigger, the centre of mass will start creeping closer to the larger mass. The best analogy of this is an athlete competing in the hammer throw at the Olympics. When the athlete rapidly spins the hammer around their body, while it might appear as though the athlete is spinning on the spot and the hammer is doing all the 'orbiting', both bodies are actually moving. There will be a 'centre of mass' between these two bodies, around which *both* are spinning. When one mass is much bigger than the other (in our example, the athlete is *much* heavier than the hammer), it appears as though the athlete is spinning on the spot and just the hammer is moving, because the centre of mass of both objects is *almost* the same as the centre of mass of just the athlete.

This same phenomenon is true with a planet and star system. The planet and star will orbit around the centre of mass of the combined system but, because the star is so massive compared to the planet, it looks like the star stays still and the planet orbits the star. But it's really important to know that it's not, and we're about to find out why.

Let's bring this all together by looking at a hypothetical star system with one planet, which you can see in Figure 10.3. To clarify, this system is not to scale. The sizes of things (the star, planet and orbits, for example) have been made much larger to demonstrate the key points clearly. In these diagrams, I want you to imagine the star in the middle, with the planet in orbit around the star, and two visible light spectra underneath. The upper, thick spectrum is what we observe on Earth (located as if we are looking at the system from Earth just as the bird's-eye view in the figure shows). The lower, thinner one is what is being emitted from the star (if unimpeded by any Doppler shifts). I've added small arrows from each spectral line of the unshifted spectrum to the red- or blue-shifted spectrum for clarity, just to highlight the subtle shifting of the spectral features. So, with that in mind, this is what happens.

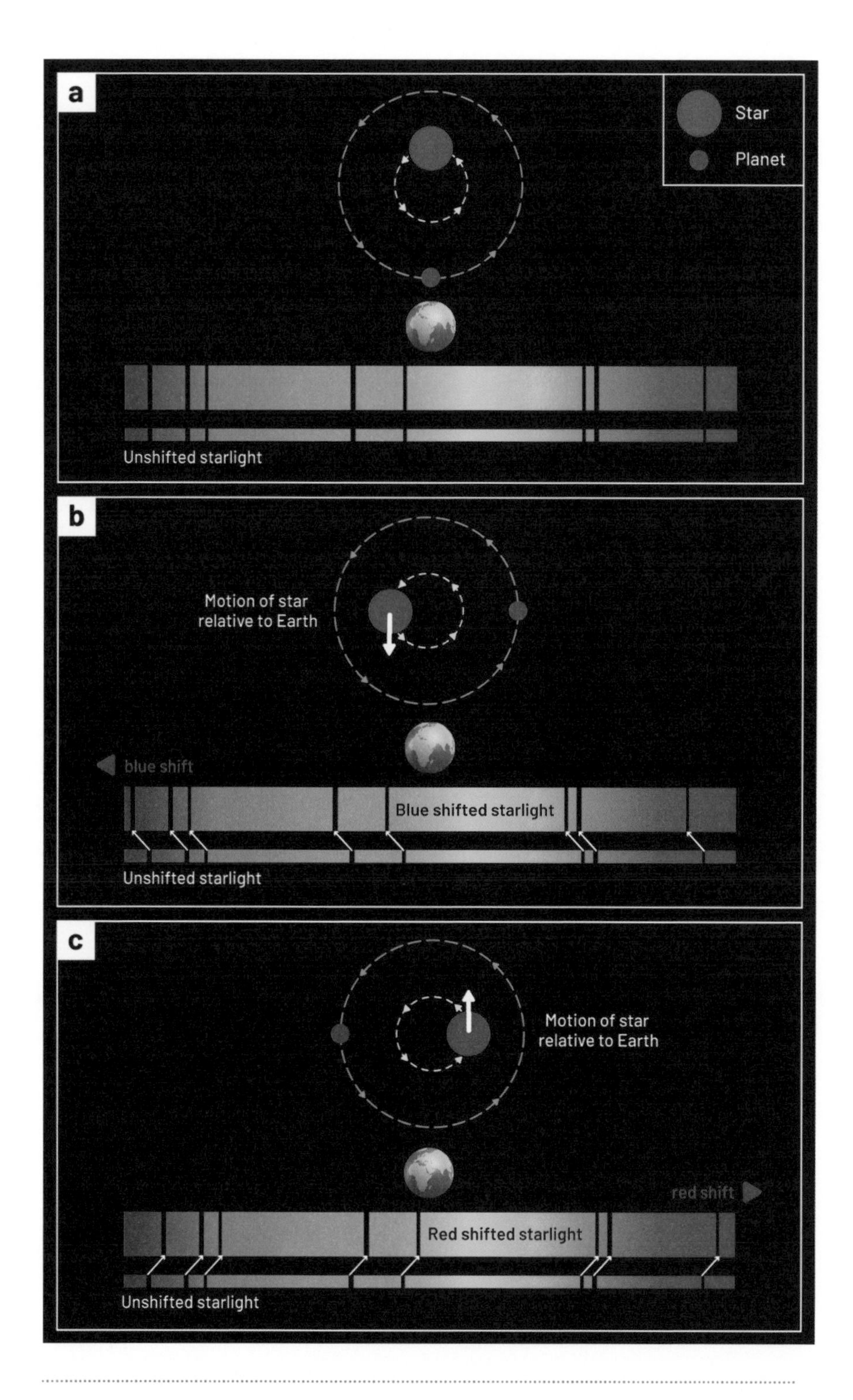

Figure 10.3: How the orbit of a planet changes the colour of the light we see of the star it is orbiting.

a It starts simply enough, with the planet and star orbiting around the centre of mass of the system, so that the star moves in a direction *sideways* relative to us viewing on Earth. As such, it doesn't squash or stretch the light wave it is emitting, so we see it just as we would expect it to be. The upper and lower spectra are aligned.

b Here is where things get interesting. The planet and star continue to orbit about the common centre of mass until the planet is orbiting one way, and the star is orbiting the other way (think about holding onto a friend's hands and twirling in a park—you'll be spinning in the same direction but, to an observer watching, one of you will be moving away and one towards them). In this scenario, the star is now moving towards Earth. As such, it is slightly squashing the light wave it's emitting, so we see it being slightly blue-shifted. You can see how the upper spectrum has shifted slightly to the blue end of the spectrum, relative to the unshifted lower spectrum (the blue versus white stars).

c Here, we have the reverse situation to (b). The planet and star continue their orbit around the common centre of mass until the star is moving away from Earth. It is now slightly stretching the light wave it's emitting, so we see it as being slightly red-shifted, relative to the unshifted lower spectrum (the red versus white stars).

You've just learned the premise of how the radial velocity method of exoplanetary detection works. Each of the physics ideas in this method is simple in isolation but, combined, they form a clever and elegant way to infer the presence of a planet orbiting another star, using sound and rigorous scientific theory.

Figure 10.4 shows how the whole process is often summarised. What is even more wonderful about the radial velocity method is that, because the luminosity (or brightness) of the star can tell us its mass, then the signal of the radial velocity can tell us all kinds of things about the planet, such as the how long the planet takes to orbit its star (which directly correlates with the distance the planet is from the star), and the mass of the planet (which directly correlates with how significantly the star is 'wiggling' or changing velocity from red- to blue-shifting. A big planet can move its star more, so the amount of red- and blue-shifting will be larger). The planet 51 Peg b that was detected using the radial velocity method is roughly half the mass of Jupiter in our own Solar System, orbiting at a distance of only about 0.05 astronomical units (about ten times closer to its star than Mercury is to the Sun) with a temperature of about 1000°C. Just let that sink in for a moment. We figured out all that information about a planet that is 50 light-years away just from applying a handful of fundamental, well-understood physics concepts.

The radial velocity method is a powerful technique, which we have used for decades to detect, and confirm, the presence of a slew of exoplanets. For a long time, the radial velocity method was *the* primary method for detecting exoplanets. It was (and still is) robust, accurate and capable of detecting smaller and more distant planets (distant in terms of how far the planet is from its host star, not from Earth) with each generational improvement in technology. Eventually, however, a new big bad boss of exoplanet discovery muscled itself to the front of the discovery statistics.

THE TRANSIT METHOD

With the help of the next generation of exoplanetary surveying spacecraft, another method (just as old, well understood and beautifully elegant as the radial velocity method) made enormous leaps forward in the number of exoplanets discovered. That technique is the transit method.

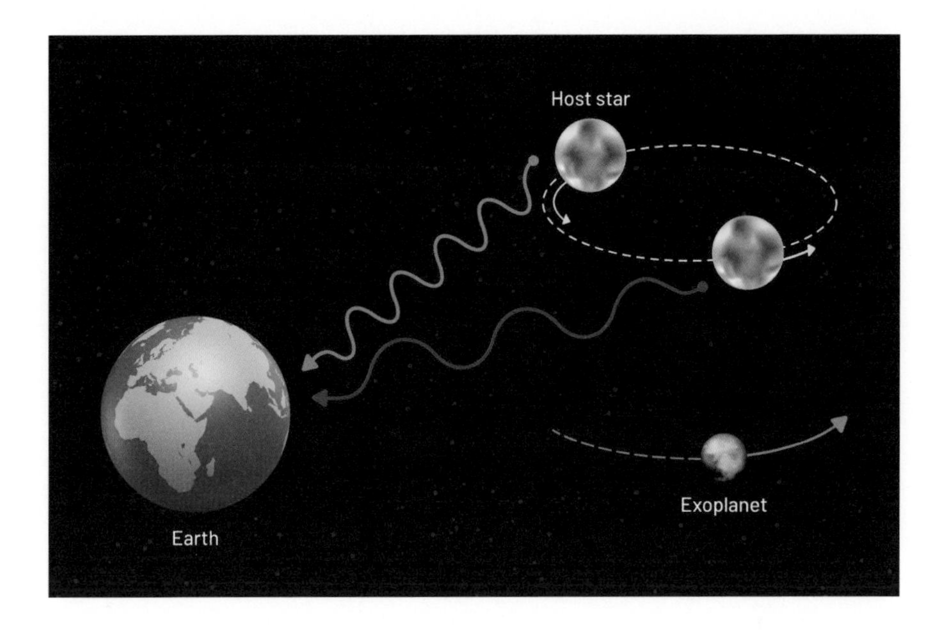

Figure 10.4: As a planet and star orbit a centre of mass, the host star goes through periods of approaching and retreating from Earth, during which the light of the star will be compressed (blue-shifted) or stretched (red-shifted), thus inferring the presence of an exoplanet.

The transit method is based around physics you may have been lucky enough to see yourself (or at the very least, you will be aware of the phenomenon). Like many terms in science, a transit has a unique definition in astrophysics, and is when a body passes between the line of sight of an observer and a body they are observing. Now, I know that sounds like a bit of a mouthful, but a simple example is a solar eclipse. During a solar eclipse, the Moon passes our line of sight with the Sun, essentially blocking out some of the Sun's light that would otherwise make it all the way to Earth. This is a transit—the Moon transiting the Sun. In a transit like this, we can't actually 'see' the Moon. What we see is the Moon blocking out sunlight. It's not invisible, it's just that what we're seeing is the absence of light where the Moon is. Combining transiting observations like this with other existing scientific understanding and context means we can declare that the huge ball in the sky blocking out the Sun is, in fact, the Moon.

As far as transits go in our Solar System, we see more than just solar eclipses. Between the Earth and the Sun are two other planets that orbit closer and nearer to the Sun: Mercury and Venus. There are times when either Mercury or Venus will cross in front of the Sun with respect to us looking at the Sun itself. Before you jump ahead and get excited about a Mercurial or Venusian eclipse, unfortunately, these aren't a thing because of how far away the planets are from Earth: they're simply too far away for us to notice with the naked eye the small amount of light they are blocking out.

When our human limitations stop us from doing something, that's when science and technology can help to solve the problem. Beautiful images have been snapped, showing exactly what it looks like when either Mercury or Venus cross in front of the Sun, thus blocking out some of the Sun's light and resulting in a clear, well-defined black circle. Figure 10.5 shows these transits, where the objects transiting the Sun are much farther away than the Moon: a Mercury transit (left), and Venus transit (right).

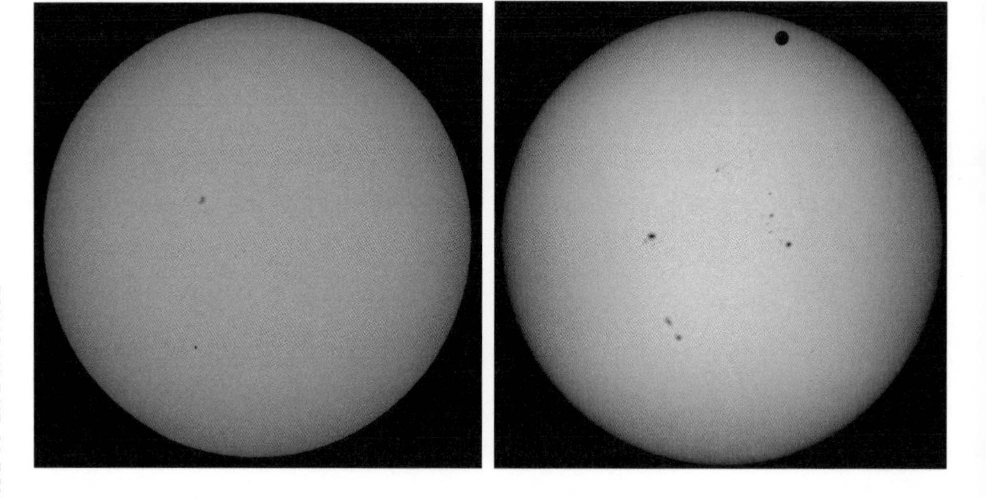

Figure 10.5: Here we can see Mercury (left) and Venus (right) transiting the Sun. Can you spot the two of them? (Ignore the blurry, discoloured sunspots—these are different 'little storms' that flare up on the Sun from time to time.)

Hard to spot, right? Mercury especially so. (The *sunspot*—that dark blob near the left-hand edge of the Sun—is bigger than Mercury!) These planets are in our own Solar System, and yet we can barely see them as they travel in front of, and block out light from, the Sun. So how could such an idea be useful for observing planets orbiting other stars that are further away? Again, we turn to technology to solve this problem; we do what our eyes do, but in a different and more accurate way.

Our eyes have loads of functions, most of which relate to our ability to interact with and interpret electromagnetic radiation (typically for humans, our eyes are most attuned to the visible light band of electromagnetic radiation). What technology allows us to do is measure the strength of light (*light* being common scientific parlance to describe electromagnetic radiation) with a much higher level of accuracy and precision, regardless of whether or not we can see the light with our own eyes.

When you're sunbathing on a lovely summer's day, you certainly don't see any ultraviolet radiation burning your skin, but we know it's there. Our eyes don't see the entire electromagnetic spectrum, but we've built technology that can. In the same way that shade from an opaque object prevents visible light and ultraviolet light from burning your skin, so does the Moon, a planet or some other small object during a transit. The object that crosses in front of the Sun acts like a wall, stopping electromagnetic light from passing through and reaching us; both the seen, and some of the unseen, forms of electromagnetic light (there's some nuance here because light diffracts around objects, but for our discussion about detecting planets, we don't need to dive into these details).

This is so important because there is no way we can *visually* see the small, black dot blocking out the light of a distant star. After all, we can barely see Mercury around our nearest star (the Sun). Regardless of whether or not we can see that phenomenon with our own eyes, it is taking place: the electromagnetic energy is still being blocked and prevented from reaching us, and that we can see! Again, not with our eyes but with our technology.

Figure 10.6 explains exactly how this works. As a planet orbits a star, at some point, the planet will cross the line of sight between us and the star (if the system we're observing is aligned so that transits can occur from our vantage point). If we were measuring the luminosity, or brightness, of the star over time, we would be able to see these small changes in brightness. We would learn how quickly the planet is orbiting the star.

During a transit, we can identify when the planet starts crossing the star (the gradual decrease in brightness), when the planet is crossing the star (constant lower brightness), and when the planet finishes crossing the star (the gradual increase in brightness). By measuring how regular these dips in brightness are, we can identify the period of the orbit (that is, how long a year is on this planet).

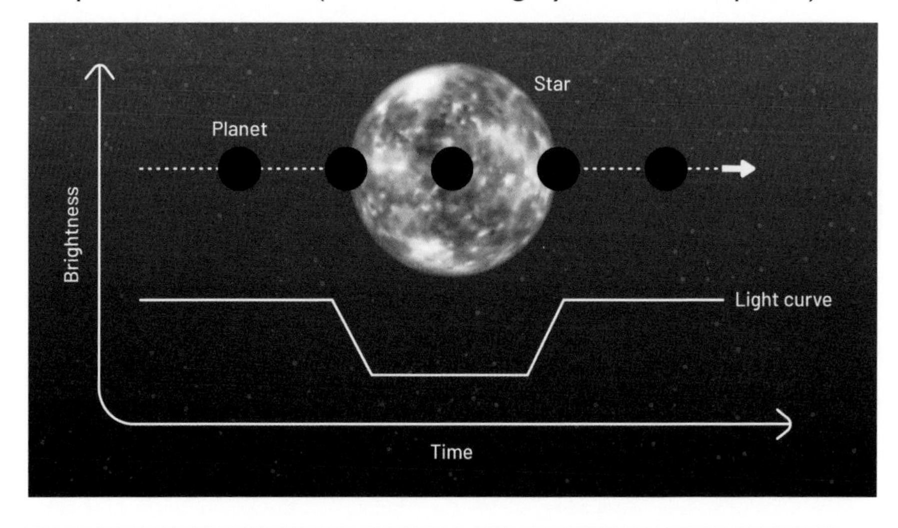

Figure 10.6: When an object transits in front of a star, it causes the brightness of the star to 'dip' as the object 'blocks out' and stops some of the light reaching us here on Earth.

Since we also know the brightness of the star (which correlates with the star's mass), we can use the mass of the star and the period of the orbit (how long the planet takes to orbit the star) to determine how close the planet is (how quickly the planet orbits its star relates to how close to the star it is, and how massive the star is). We can also determine, based on how much the light drops, how big the planet is (the size, as in how much light is being blocked out, not necessarily the mass, but sometimes we can use clever ways to make reasonable estimates).

If you thought the radial velocity method was beautiful in its simplicity, then the transit method is pure, unapologetic elegance. The physics, and even just the fundamental principles behind the idea, are delightfully unostentatious. I remember two things vividly when learning about it for the first time. First, my mind was blown away. Detecting planets around other stars! Second, and more introspectively, I was almost slightly amused and humbled that we hadn't been doing something like this sooner.

To me, this is just another of the many beautiful parts of science. There are breakthroughs and findings of such sophistication and complexity that you need to dedicate decades, if not your life's career, to honing your craft and adding to our (humanity's) collective body of knowledge. But there are still some that—like most brilliant inventions or innovations—can almost be mistaken for being a bemusing oversight at first glance. Both these types of breakthroughs are equally valuable, and both can lead to profound and rewarding outcomes, which can turn our understanding of the world on its head.

Science is a far-reaching beast that offers opportunities to people of many different interests, abilities, career aspirations and—let's be fair—affinities for luck. It's a gentle reminder that science is not only for the elites, for the Newtons or the Einsteins, or for those who achieve things that get named after them. No, science is for everyone!

I've well and truly digressed here though, so let's jump back to the transit method.

Wow, right? What a brilliant way to detect planets. It's been such a successful method of exoplanet detection that it's become the primary means of detecting planets outside our Solar System. The transit method, with the help of the latest two exoplanet survey spacecraft—the *Kepler space telescope* and *Transiting Exoplanet Survey Satellite* (TESS)—has seen an explosion of exoplanet discoveries, such that now (as of March 2022), we have confirmed the existence of more than 5000 exoplanets (confirmed meaning they've been validated by multiple observations), as well as more than 7400 more potential candidates (of which most are likely to be planets and not false positives), making for a mix of more than 10,000 confirmed and candidate exoplanets, spanning across more than 3000 planetary systems. That is mind-boggling, considering it was only back in 1995 that we found the first planet outside our own Solar System (that is, a second planetary system). Only a few decades later, we're at 3280 planetary systems. What an enormous and rapid leap! The transit and radial velocity methods are not the only ways to detect exoplanets, but they make up approximately 95% of all discovered exoplanets, so to say the transit and radial velocity methods are successful is a huge understatement. Together, they have opened the door to a new world of exoplanetary science.

Exoplanet observations don't just allow for planets to be detected, but for us to glean more details about them, such as their size, location (relative to the star) and, in some cases, chemical composition and atmospheric details. Exoplanetary science is moving so quickly that new and more exciting possibilities are becoming a reality at an ever-increasing speed, which is helping us to continue our search for Earth 2.0 and life beyond Earth.

In the next chapter, we are going to look in detail at some of the properties of exoplanets we can determine when we observe them, what types of exoplanets we have found, and what we can learn about the atmospheres of these distant worlds.

CHAPTER 11

FLAVOURS OF EXOPLANETS

Planets orbiting a system beyond our own star—what an exciting idea! Just as the planets in our Solar System orbit the Sun like clockwork, so too do exoplanets, but they're orbiting a completely *different* star. It's a powerful notion when you think about it.

Some of these planets would enjoy the beautiful dance of the morning star-rise, and equally playful aesthetics of the fleeting star-set. (Sensationally spectacular to be sure, but notice our use of 'star-rise' and 'star-set' as opposed to sunrise and sunset. These are happening on planets orbiting other stars, too.) All the breath-taking landscapes and scenery, escarps and outcrops, oceans and lakes; these are at the mercy of the star the planet orbits; they're bathed in the electromagnetic rays being emitted from the star, not just in the form of beautiful visible light during star-rise or star-set, but often the dangerous invisible portions of the electromagnetic spectrum.

As well as all this light, these planets are also blasted by highly energised and ionised stellar ions (we use *stellar* instead of *solar* here, because we aren't referring to our own star, *Sol*). And just like the Earth, some of these planets would be exposed to the radiation and ions their host star emits. Space is dangerous—stars particularly so—and as we discussed at length in Part 1, through a series of fortuitous circumstances, we've found ourselves on the only known planet that has, in some way or another, ended up as *the* perfect location for life to not only begin but thrive. It's protected and stable, and it offers a safe haven within which life prospers.

INVESTIGATING EXOPLANETS

When we consider exoplanets, and the properties and attributes that make them what they are, there's a lot to think about. The vast distances these exoplanets are from Earth makes it particularly challenging to glean a rich, vibrant understanding of them. In a testament to some of the most quintessential human traits—our innovation and ingenuity—we've come up with clever techniques and approaches that we employ to better understand details about the nature of the planetary systems we observe beyond our own Solar System.

In this chapter, we're going to expand on some of the exploratory techniques we covered in Chapter 10, and how we use these methods to extract as much information out of an exoplanet as we can. We'll also start looking at some of the weird and extraordinary worlds we've found so far in these early stages of exoplanetary science.

We discussed in Chapter 10 how the transit and radial velocity methods are critical for observing (or inferring) the existence of an exoplanet, but these methods are also really useful to help us better understand the characteristics of these exoplanets, despite their distance from Earth. When we detect an exoplanet, a flurry of questions come to mind. Is it small and rocky like Mars? Large and icy like Neptune? Huge and gassy like Jupiter? Could it have an atmosphere? What about all the other things that might exist on these planets that don't exist here on Earth? These are all important questions, and some we can hopefully answer, or work through to get a greater understanding.

So, we're looking at a star. We've got at least two ways we can detect planets around a star outside our Solar System and ... Is that it? Is that where the story stops? We found planets, but who knows what they look like? Fear not, I wouldn't lead you down the garden path only to disappoint you like this. (Although, cards on the table, I will *sort of* have to at some point—but I'll do it as gently as I can, I promise!) No, we can learn quite a lot about exoplanets, using several clever observations, comparisons and scientific analysis.

HOW STARS GIVE UP THEIR SECRETS

Let's recap a couple of important things from Chapter 10 about stars. First, their brightness is directly correlated with mass. The brightest stars really do burn out the fastest, and the main reason for that is all to do with our good friend gravity. Essentially, the bigger and more massive a star is, the more gravity there is, squashing and crushing it all together. Cast your mind all the way back to Chapter 1; you'll recall that, once there is enough stuff squeezed into a small-enough space, nuclear fusion commences, which is the combustion engine of our star automobile. The material is squashed so tightly together it ends up combining, or fusing, and because of some *very* important properties of science, an enormous amount of energy is released when this happens.

So, how does the size of the star fit into all this? Remember in the previous chapter when we talked about basketballs bouncing around the stadium or bedroom? The bigger and more massive a star is, the more gravity is pulling and squashing everything together, and the more often these nuclear fusion reactions take place, thus releasing energy more rapidly. The basketballs squashed into the tiny bedroom collide more often than the basketballs in the stadium. This means a bigger star will force things into a smaller space, and create more fusion reactions more quickly (making the star appear brighter), but if the fusion reactions are quicker, that also means it will cause the star to run out of fuel more quickly.

Big bright stars burn the brightest (and die the quickest), and that is an important relationship for us to have established. Since we can actually measure a star's brightness accurately, and because of the well-understood relationship between brightness and mass, we can determine the mass of the star by observing its brightness. If we add the temperature of the star into the mix (something we can infer by the colour of a star), we can go another step further and estimate the radius of the star as well. From what looks like a mere pin prick in the sky, we can obtain a great deal of detail about the properties of that star, including its brightness, mass, size (radius) and temperature. Once we know all these details about a star where exoplanets are orbiting, the radial velocity and transit methods grant us an avenue to learn more about these exoplanets, too (even though the exoplanets are smaller and fainter than the pin pricks in the sky that they themselves are orbiting about!).

There's plenty to unpack here, so I'm going to do it reasonably methodically: first, what the radial velocity method teaches us about planets; second, what the transit teaches us about planets; and finally, what both of them (often in combination) teach us about planets.

WHAT THE RADIAL VELOCITY METHOD TEACHES US

In the last chapter, we discussed our hammer-throwing athlete, and how that was a good analogy for Earth orbiting the Sun: our planet (the hammer) moves *a lot*, whereas our Sun (the athlete) moves very little. To a distant observer, it may look as though the athlete isn't moving at all. This part all makes sense to us, but now let's take it to the next step—figuring out how the size of the 'wobble' of the star can tell us about the size of the planet. Just as a quick aside, I'm going to use the word 'wobble' a lot when I talk about the star orbiting about the system's centre of mass (the system being the one that the star shares with its other planet/s) because I think wobble is a really good word for it. The gravitational tug-of-war between the planet and star means the star doesn't get pulled around nearly as violently as the planet does. It does gets nudged, however; it does wobble a bit, and, fortunately for us, our technology is fantastic at detecting these subtle wobbles.

We talked about how we detect these stellar wobbles in Figure 10.3, using a series of diagrams with colourful shifting spectra (the rainbows with spectral features). This mechanic is critical to understand not only that a star *is* wobbling (from this wobble, we can infer the presence of the planet) but also *how much* it is wobbling. Here is where the correlation between stellar luminosity and mass becomes significant. Depending on how bright the star that we're observing is, we also know how massive (or heavy) it is, because of the well-understood relationship between brightness and mass. Because we know how heavy this star is, and we have observed *how much* it is wobbling, we can begin to discern exactly how massive the planet must be.

Remember our analogy about the hammer-throwing athlete? When spinning a much lighter hammer, the athlete barely moves at all. Alternatively, if the athlete picks up their child and spins them around in a playful manner, then the athlete's entire body also spins to maintain balance. In both these scenarios, the athlete (the star) has not changed. They just have to move, rotate and wobble more when spinning their child to maintain balance because their child is more massive than the hammer. That's the exact same premise as what's happening here with the radial velocity method. If the star is moving by only tiny amounts, then the orbiting planet must be pretty small, whereas if the star is wobbling and dancing around significantly, the orbiting planet must be pretty big. There's a little more nuance to it than just this (especially regarding the distances between the star and planet) but we'll bring it all together later, so don't worry too much about it right now.

How cool is that though? Let it *really* sink in. We can look at a tiny light in the night sky and understand that it's a raging, living star like our own Sun; we determine the mass, temperature and radius of that star; we can detect whether that star is wobbling due to the presence of a planet in orbit around it; and we can determine how massive that planet is according to how much the star is wobbling. What an idea to entertain, let alone witness it play out right in front of our own eyes!

WHAT THE TRANSIT METHOD TEACHES US

We're on a roll here now, so let's jump straight into the transit method and see what we can learn about the planet using this technique. Well, there's something elegantly simple right there in front of us, and it has to do with the planet's size. This time, though, I truly mean its size—its dimensions; the planet's width or diameter—not mass, or how heavy it is. The planet's size is an important detail during a transit because that determines how much light is going to blocked from reaching us here on Earth. A rocky little friend like Mercury passing in front of the Sun? It'll block out very little sunlight. It's tiny compared to the Sun. Mercury is so small that you could line up 285 Mercuries next to each other and they still wouldn't quite be the same width as the Sun. So, when a small planet like Mercury transits the Sun, not much light is blocked at all. In fact, if an observer outside our Solar System measured a Mercury transit, 99.999% of all the Sun's light would still reach them. Mercury doesn't make much of a dent in how much light is observed.

Let's compare that to a big planet, like Jupiter. We know from Chapter 4 that Jupiter is the big sucker of the Solar System. It's so big that you only need to line up ten Jupiters back-to-back and it's almost bang on the width of our Sun. So how much light would an observer outside our Solar System measure when the biggest, baddest planet in our Solar System is blocking out some of the light? It's a little under 99% (98.992% if you want to be more precise). I imagine your first thought is, 'Isn't 98.992% pretty similar to 99.999%?', and you'd be both right and wrong. They're both close to 99% if you only count that much precision, but if you include all those extra digits (which you must when dealing with some precise science like this) the difference ends up resulting in Jupiter blocking out over 1000 times more light than Mercury. I imagine then your second thought is that it's still blocking out a tiny fraction of light, and you'd be right. Observing from outside our Solar System, Jupiter would only block out about 1% of the Sun's light. And that's sort of the point: planets are big, but stars are enormous, so we need to measure the brightness with sophisticated and highly calibrated telescopes to gain the level of accuracy and precision we're after.

But don't let that number disappoint you either. Just think about what it means for a moment: a planet blocked out more than 1% of the light of its star (if 98.992% was observed then 1.008% was blocked). Considering the host star is the heart, soul and foundation upon which a planetary system is built, that one of its smaller planetary companions can block out an entire per cent of light is no small feat.

THEIR POWERS COMBINED

These two exoplanet detection techniques are not only wildly successful in finding planets outside our Solar System, but are also excellent at providing us with additional detail about the planets themselves—detail and information we may never have thought possible.

What else can we learn from these two techniques? What about one of the properties that can be determined from both: the distance from the star, or more specifically, the size of the planets orbiting it. In both methods—the star's circular wobble as the planet orbits around it, and the star's cyclical dropping in brightness as the planet crosses in front of it—happen like clockwork.

This means that we can determine how long a year is for that planet. What if the light of the star is blocked out every 217 days? Looks like the planet takes 217 days to orbit the star. Likewise, if the alternating red- and blue-shift of the star occurs every 217 days, then it also means the planet is orbiting every 217 days. The regularity of the techniques lend themselves to identifying the length of the planet's year, or what is referred to as the *orbital period*. But how do we work out how big or wide this orbit is?

This is where our good friend gravity helps out yet again. As we've mentioned time and time again, the closer things are to massive objects, the harder gravity acts on them. The increased gravity nearer to a star causes the planet to orbit faster. Not only is the planet hurtling through space faster because of this stronger gravity, but the size of the orbit is smaller; it has far less distance to travel to complete a lap of the star. These two facts combine so that orbital periods, or years, of planets are shorter for those planets nearer to their star. (It takes Mercury only 88 days to complete an orbit, compared to 365.25 days for Earth, 4,332.6 days for Jupiter and more than 60,000 days for Neptune!)

The only problem is that there are many ways to achieve a particular orbital period. Is the planet on a small orbit around a low-mass star? Or is the planet on a fast orbit around a high-mass star? Or something in-between? The last piece of the puzzle—which we conveniently identified earlier when talking about the brightness and mass of a star being inextricably linked—is the mass of the star. If we've got the length of the year (orbital period), and the mass of the star, we can determine how close the planet must be to have an orbital period equal to what we observe. So now, we not only know the mass and size of the planet, but we know how close it is to the star, too.

The final piece of the puzzle combines elements of both the radial and transit techniques to determine the chemical composition of the planet. The way it works is that, from the radial velocity technique, we know the planet's mass, and from the transit technique, we know the planet's size (its physical dimensions). By combining these two terms, we can estimate the planet's density. This is written as an equation like so:

$$\text{Density} = \frac{\text{Mass}}{\text{Volume}}$$

What this means is that the density (p pronounced 'rho') is equal to mass divided by volume. Although we don't know the volume, we do know the width (or diameter) of the planet, which we can use to estimate the volume, assuming it's a perfect sphere. (This isn't an unreasonable assumption—remember hydrostatic equilibrium? As things get bigger, nature tends to want to form a sphere.) While there are some caveats and assumptions that limit this, it's a great starting point because when we discover planets around stars, we can estimate the planet's density to get an idea of whether it's made of lots of dense material (rocky), lighter material (gassy) or something in between (water), and thereby discern what type of planet it is (because we know from within our own Solar System there's a lot of variation among planets).

TYPES OF EXOPLANETS

Knowing an exoplanet's mass, size and density allows us to come up with a starting point of what 'types' of planets we're looking for beyond our world. Let's have a look at some here:

1. **Rocky planet** This is pretty self-explanatory: an object that's large enough to be a planet, but small enough to be a rocky planet (as opposed to an ice or gas giant). In our Solar System, Mercury, Venus and Mars are rocky planets.

2. **Earth** Well, this is the goal, right? It's technically a rocky planet, but it's in its own classification because of how significant it is (we also tend to consider its location). A planet similar in size to Earth, which is rocky and in that sweet Habitable Zone region around the star.

3. **Super-Earth** This type of planet doesn't exist in our own Solar System, but with more observations, it's becoming clear that it's potentially the most common type of planet, and so it's actually a bit of an oddity that we don't have our own super-Earth. As the name suggests, this planet is like Earth but bigger, usually up to twice as wide (which can mean eight times as massive). We don't know loads about super-Earths, other than they're apparently common but, for some reason or another, there isn't one in our Solar System. If found in the Habitable Zone, would a super-Earth be just as ideal for life as a regular-sized Earth? Or is there something that makes it less than ideal? We're not so sure yet, all we know is that there are a lot of them, so still plenty of questions to be answered.

4. **Ice giant** This is the second-biggest type of planet, and is aptly named for having surface temperatures of about –200°C. Uranus and Neptune are our ice giants. They have interesting chemical compositions, and they're remarkable from a planetary formation and Solar System history context. But for the

purpose of our search on better understanding life beyond Earth, we'll move right along.

5. **Gas giant** This type of planet is the big sucker in a star system. In the Solar System, Jupiter and Saturn are not only the biggest but, arguably, two of the jewels in its crown (I'm thinking of Saturn's rings, of course, but Jupiter's storms are equally mesmerising). These two play an integral role in planetary formation but, as we discussed in Part 1, they may also be important by being benevolent protectors against a barrage of life-extinguishing asteroids. The gas giants may be quite important for life to exist.

6. **Hot Jupiter** This is a special category of planet, which I wanted to include because I love the name (who doesn't?) and because they were the focus of my Masters' thesis, so I have a soft spot for them. Hot Jupiters are essentially Jupiter-sized planets that, for some bizarre reason, have ended up orbiting closer to their star than Mercury orbits the Sun. Remember, Mercury's year is 88 days, but that's far too slow for hot Jupiters. And they're so close to their stars that they're orbiting in mere days! And they're as big as Jupiter! Somehow, some enormous Jupiter-sized planet has muscled in so close to its star that it completes a lap around it every couple of days. (Imagine if one year was only a few days!) They're just weird planets. It's weird that they're so close to their stars. It's weird how they formed. But they're fascinating because they're incredibly hot (being so close to their stars, these planets are often several thousand degrees hot) and have well-inflated atmospheres that make their atmospheres much easier to study (we'll get to that shortly). In terms of exoplanetary science and understanding exoplanet atmospheres, they're like little floating laboratories for us to use.

There's certainly significant variation in the types of planets we've observed. Fortunately, we have a few types here in the Solar System, and *most importantly*, the one that produced life. But wouldn't it have been great to have a Super-Earth in our Solar System to study as well, to understand the nature of these kinds of planets in greater detail? They seem to be the most common type of planet—at least, in our little corner of the Milky Way where we have been searching—yet, rather oddly, we don't have one in our Solar System. Why is that? Maybe more importantly, though, is the question: other than their size, how similar to Earth are they? Are they even like Earth at all? Are they potential worlds to explore for life? Or is there something bubbling under the surface that changes the nature of the planet itself? Right now, we just don't know.

ILLUMINATING ATMOSPHERES

There's one more super gnarly aspect of exoplanet properties that I *simply* must share. I vividly recall sitting in my Masters' supervisor's office, my mouth agape, as he was explaining it all to me. Just like so much of this stuff, it seems deceptively simple, yet it's profoundly impactful. Observing a star before, during and after a planet passes by can provide us with so much information about what's going on in the atmosphere.

Check out what's going on in Figure 11.1, which shows what happens when a planet orbits in front of the star we're observing (called the *primary eclipse*). You'll notice there's a blue ring around the planet. *That* is the planet's atmosphere. What's crazy to think about is that the star's light shining through this atmosphere will interact in such a way that we can learn about the exoplanet's atmosphere all the way from Earth. These planets are so far away that the idea of 'seeing' one is completely fanciful. And yet, using fundamental physics concepts and cutting-edge technology, we can learn about a planet's atmosphere in a system far, far away. And that's truly astonishing.

Figure 11.1: In a primary eclipse, a planet passes in front of the star we're observing, causing the star's light to shine through the atmosphere and illuminate it.

To completely understand this, Figure 11.2 contains images to show how this works. Right now, this technique is being used to study and understand the atmospheres of planets we can't even see; planets we only know exist because of how they interact with the stars they orbit. Essentially, during the primary eclipse (where star light shines through the planet's atmosphere and reaches us on Earth), two scenarios can happen.

a **The simplest scenario** When a planet has no atmosphere or a very thin atmosphere, during an eclipse, the light still passes straight through to Earth unimpeded. To be sure, the planet has blocked out some of the light, so we can still learn things about the planet, regarding its size, the period of its orbit (length of its year) and its density and planet type, but we can't learn anything about the planet's atmosphere (if it has one!).

b **Things start to get interesting** When the planet has an atmosphere that is thick enough for the star's light to flow through, all sorts of things start to happen. Basically, the electromagnetic spectrum (which we've discussed many times before) will interact with chemicals in this planet's atmosphere in one of two ways. Either the light is scattered—it hits one chemical and sort of gets re-routed in a different direction (typically not in the direction of our Earth) so we lose that piece of light, or it is absorbed.

We've discussed this scenario before with our dancing eggs, where certain chemicals absorb certain colours of light to dance. When this happens, the absorbed colours are prevented from reaching us on Earth. In both cases—whether the atmosphere scatters or absorbs

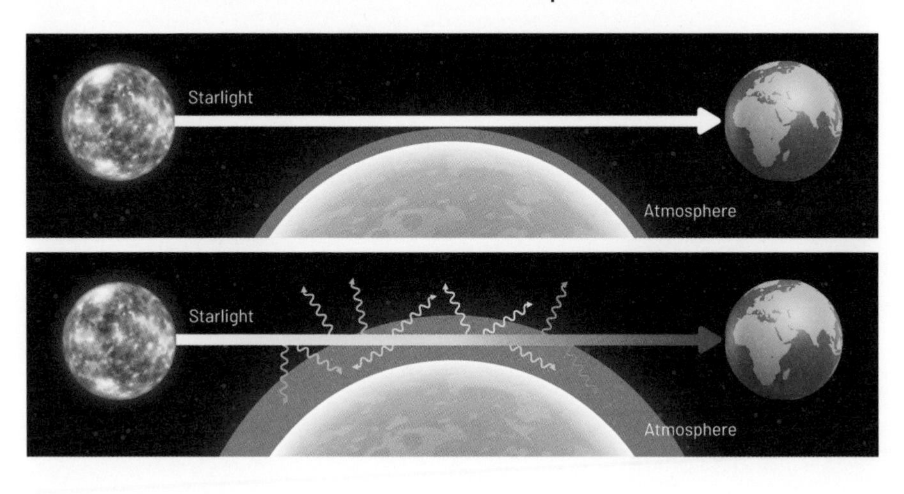

Figure 11.2: A diagrammatic representation of how the light from a star interacts with the atmosphere of an orbiting planet before it reaches Earth.

the light—these physical processes interact with certain colours of the star's light and prevent them from reaching us on Earth. Using those missing colours in the spectrum—the signature or fingerprint—we can deduce what the chemistry must be in the planet's atmosphere.

We can be observing a planet that we can't even see and yet we can figure out (with caveats and limitations, of course) what's going on in that planet's atmosphere. What chemicals are there? What does the structure of the atmosphere look like? Does it form clouds?

Does it rain? These are questions and areas of astrophysics being researched today. That we can already gauge the atmospheres of planets dozens of light-years away shows that, with each generational improvement of technology, we're one step closer to answering some of these questions with more confidence and deeper understanding.

We've reached a point where we could actually find a planet similar to Earth and start learning about the chemistry of its atmosphere—that's pretty incredible and it's exciting to see what we will find next.

From our discussions in Chapters 2 and 7, being able to observe an atmosphere on these distant worlds is really important. Not just from the perspective of the whole 'How good is breathing?' angle, but also from the 'protective sleeve' angle. Remember, to hold down liquid water, we need a sleeve, which can be in the form of an icy shell or, preferably, an atmosphere. Because we can now observe planets and identify if they have atmospheres, we're another step closer to finding another place like Earth or even (scarily?) another place with life.

So, where to from here?

We've looked at why the Earth is special, and where else is special (both within our own Solar System and beyond). It's starting to look like there are quite a lot of special places, even with our current limitations in finding them. With that in mind, it's probably time we start thinking about some of the really big questions.

Where are the aliens? How are we looking for them? And why would they possibly be hiding?

PART THREE

WHERE ARE THE ALIENS?

What a wonderful, perfectly balanced and delightfully mysterious Universe we find ourselves in. We've looked at some of the most wonderful aspects of Earth—things that make it striking and unique—not only from biological and ecological points of view, but maybe even more profoundly, from an artistic perspective. Over the span of millions of years, we have developed an ability to contemplate and assess something by how aesthetic, beautiful and artistic it is—which is quite odd if we stop and think about it. Somehow, the relentless march of natural selection has led us to develop this ability to appreciate the beauty of the world around us. It is both a delight and a puzzling conundrum that we can recognise the magnificence of Earth by the intricate combination of its visual beauty and its unique position in the Universe as the only known place to support life.

But just how unique is Earth? Are we *really* the only place to support life?

When it comes to the reality of our search for life beyond Earth, Arthur C. Clarke says it best: 'Two possibilities exist: either we are alone in the Universe or we are not. Both are equally terrifying.'

This notion is not only insightful, terrifying and exciting, but also philosophical, because it acknowledges that the answer to Are we alone? would fundamentally change our understanding of the world around us. He has highlighted why this area of research commonly comes up in astronomical circles as one of the biggest questions in astronomy right now. The bluntness of the quote leaves us to contemplate just how profound both circumstances really are: Surely, we're not alone. But what if we are?

(Spoiler: this book can't give you the answers you seek. We're still figuring out whether or not we're alone. But, hopefully, it will provide some context as to why our pursuit of life beyond Earth is important: we're getting closer to uncovering answers to questions that will have an enormous impact on our understanding of where we fit in the Universe.)

We're going to delve into some of the more 'out there' stuff. Aliens, their existence, where they are, have we seen them an, if not, why not? We're going to traverse these questions while remaining firmly grounded in science, with just a pinch of creative licence.

Because that creative licence is what inspires the next generation of scientists. It's what lights the fire of curiosity to take the research as it exists today and ask the question: What about *this* instead? This is where real science can often happen, when we entertain 'outside the box' thinking—radical, fun and wild ideas. So, enjoy Part 3. Let it blow your mind. Let it wow you, let it scare you but, ultimately, let it take you on a journey you'll enjoy. Hopefully by the end, we'll have the context and tools to think about and critically discuss answers to the question: Where are the aliens?

CHAPTER 12

HOW COMMON?

Let's just tell it like it is: we know only one place where life exists, and that's right here on Earth. We've spent Part 1 of this book going through the wonders of Earth and what makes it special: the existence of water at just the right conditions, an atmosphere, a protective magnetosphere, a (relatively) large Moon, the presence of Jupiter and plate tectonics.

We then spent Part 2 learning that, actually, there are plenty of other places in the Solar System that are pretty special, too. Maybe not *as* 'special' to us because they may have subsurface liquid oceans trapped under kilometres of ice (not our ideal living circumstances!), but as far as life goes, they're pretty special locations. We looked at three of the most interesting locations in our own Solar System from a liquid water and possibility of life perspective.

In the second half of Part 2, we looked at planets beyond our own Solar System, so-called exoplanets. We're finding them not just in the thousands but in the tens of thousands. You must be thinking, surely, it's no longer about *if* we will find evidence of alien life, but *when*.

Our galaxy, the Milky Way, is teaming with hundreds of billions of stars, and we've only searched for exoplanets around approximately a million stars if we're being really generous (as of 2022). This is a mere fraction of a percentage of the stars in the Milky Way, and yet within that small amount we've uncovered several thousand planetary systems (with our current technological limitations). The numbers are stacked in favour of there being life somewhere else, because the alternative just seems astronomically unlikely (pun forcefully intended) to be almost impossible. There must be other places where life has existed, survived, thrived and then (rather grimly) likely perished. So where do we start trying to wrap our heads around a subject that is, of course, not only profound, but also has little real data to work with?

THE DRAKE EQUATION

Dr Frank Drake put forward one way to tackle this, with the aptly named *Drake equation*. It uses a probabilistic approach to searching for life beyond Earth. This is where we want to start our discussions on Where are the aliens?—it's a way in which we can quantify our search. It's tricky to talk about the Drake equation for so long without seeing the equation itself, so without further ado, let's have a look at this beast in Figure 12.1!

$$N = R_* \times f_p \times n_e \times f_l \times f_i \times f_c \times L$$

Figure 12.1: The Drake Equation

Eight letters. That's the Drake equation. Delightfully simple, yet devilishly impactful in its implications. It's time to start unpacking this! There's a whole lot of multiplication taking place on the right-hand side of the equation, so let's start with the easier left-hand side.

THOSE ARE SOME BIG NUMBERS!

One of the crazy things about large numbers is that we just can't get our heads around them. When something is beyond our immediate ability to see, use or understand it in a tactile or visual way—it's referred to as being abstract. The problem we have with mind-bendingly large numbers, such as that there are about 100 billion trillion (yes, with a 'b' and a 't'—a 1 followed by 23 zeroes!) stars in the Universe, is that they're so big, they're almost meaningless.

It's more than the number of synapses in the human brain, and it's really hard to comprehend and understand. We often rely heavily on metaphors and analogies to help us to contextualise these sorts of things and illustrate just how unfathomably large these numbers are. A common analogy for the number of stars in the Universe is one that drives some people to utter disbelief—and that's sort of the point!

The analogy is that, for each grain of sand on Earth, there are 10,000 stars in the entire Universe. Every grain of sand—between your toes on the beach, being churned by the crashing waves on the beach, being carried gently by one of the many rivers flowing across Earth—all of them, each one, is the equivalent of 10,000 stars in the Universe. It's not just incredible to think about how much sand there is on Earth, but it also really hammers home just how large the Universe is, and how many stars really are out there.

N

This is the part of the equation we're trying to calculate. What is the number of civilisations in our own galaxy (the Milky Way) that are *communicatively intelligent?* I really do want to emphasise the 'communicatively' part of this because the key goal of the Drake equation is to quantify the galactic civilisations we could actually communicate with. This is a subtle difference from the number of 'intelligent civilisations', but it's critically important, as we'll learn later when we get to the ability to communicate.

For now, let's examine the right-hand side of the equation.

R_*

The R_* in the Drake equation is the rate at which stars are being created in our own galaxy, or to put it another way: What is the Milky Way's birth rate? This is going to change over time—younger galaxies are much more fecund regarding their star birth rates. As of right now, the Milky Way's birth rate is equivalent to the mass of 1.5 Suns per year. This does *not* mean the Milky Way is pumping out 1.5 Suns per year, however, but the amount of material it's collapsing together and turning into stars is equal to that of 1.5 Suns. On average, stars tend to be a little bit smaller than our own Sun (about half as much mass), so a *mass rate* of about 1.5 Suns per year is the equivalent to a *number rate* (that is, how many stars are formed) of about 3 new stars per year.

The f_p in the Drake equation represents the fraction of stars with planetary systems. If $f_p = 1$, for instance, then every single star has a planetary system, but it doesn't tell us anything about the planetary system: How many planets are there? What type of planets are they? Are there big gas giants like Jupiter? *Are there Earths? Conversely,* $f_p = 0$ means that no star has a planetary system, something we know to be categorically false (because we exist). So, we know f_p must be greater than 0 but that's all we know with certainty.

It's most likely to be nearer to 1 than to 0, because it's expected that most stars will have formed with planets (due to the nature of how stars form: after they form, typically, there's a massive disc of material still orbiting the star, from which the planets form). There are always exceptions, as well as situations where systems passing one another can result in planets being ejected from either system, leading to stars with no planets left (and planets with no star, so-called rogue planets). So, while 1 is the hypothetical maximum limit, it is an ideal scenario, and likely not realistic. Complications will likely be rare, however; so the number is more likely than not going to be a fraction very near to 1 (think somewhere in the 0.9 to 0.99 range; a case of the rule being that a star has a planet, and the exception being a star not having a planet).

n_e

The next part of the Drake equation doesn't start with an *R* (that is, it's not a rate) or an f (that is, it's not a fraction), but an *n*, so what is it? It's a bit of a mouthful, actually—it's the average number of planets (per planetary system) with an environment we would consider suitable and capable of sustaining life. Right now, in the Solar System, $n_e = 1$, that is, there is 1 planet with an environment we would consider suitable and capable of sustaining life (the Earth). In the past, however, perhaps there were times when the conditions on Mars, and indeed Venus, were very different. Maybe, before the atmosphere of Mars had all but disappeared, leaving the planet's liquid water to be either frozen or evaporated; maybe, before the rampant, run-away greenhouse effect on Venus scorched the planet and left it as the hellscape it is, buried under the crushing depths of its thick opaque atmosphere; maybe, before these events (that took place over hundreds of millions of years ago) these planets were also considered cradles of life; capable of birthing life.

Maybe, at some stage in the past, n_e = 3 for our Solar System (the Earth, as well as the Mars and Venus of old). But this changing or evolving number doesn't matter when we're dealing with such a large sample (that is, the hundreds of billions of systems we're considering in the Milky Way). The idea is that n_e is the average of how many planets capable of supporting life exist in any planetary system. The current estimate is about 20% or one-fifth, so let's put it another way: for every five planetary systems, we could expect one of them to have a planet like Earth.

f_l

In the Drake equation, f_l represents the fraction of planets capable of supporting life that *actually develop at some point.* There is a subtle difference between the number n_e and this fraction f_l. Whereas n_e indicates what proportion of planetary systems is likely to have a planet like Earth that is *capable* of sustaining life, f_l is the actual fraction in which life *actually manifests.* The environment of a planet can be all ready to go—perfectly warm and moist, a protective atmosphere and magnetosphere, adequate but not too abundant vulcanism, and plate tectonics—an entire smorgasbord of special ingredients, ready to dance together and allow for life to coalesce and go forth. But sometimes, it doesn't. Maybe f_l is the fraction that incorporates that sort of potential 'missing link' between the ingredients for life and life itself starting. Is there still some special event or catalyst that kickstarts the process, which we don't know about?

So f_l = 1 would mean that, if an Earth-like planet exists, life will develop as a sort of inevitable natural part of evolution. We don't really know much about what this number could be, other than that it's certainly above 0 (again—because we exist) but research is currently favouring the hypothesis that it's likely a high fraction near to 1. It's a case of combining all the right ingredients for life on a long enough timeline, and life will begin.

This is where we start getting into the really juicy stuff: f_i represents the fraction of planets that have developed intelligent life, as opposed to merely life itself. Remember, these fractions are independent of each other—multiplying the whole equation at the end is what combines it all together to give us N (the number of civilisations in our own galaxy, the Milky Way, that are communicatively intelligent).

So, when we're talking about f_i we mean the fraction of those planets where life has already popped up. It's like saying: 'In these 100 houses *where life has developed*, the fraction f_i will be the number we can talk to (that is, are intelligent).' Whereas previous fraction f_l is like saying: 'In these 100 empty houses, the fraction f_l will be the number where life develops.'

Notice the subtle difference? This also hints at how we will be able to combine these numbers multiplicatively to answer a broader question: 'In these 100 houses, what fraction have life we can talk to?' We don't have a great estimate here, and there are arguments going both ways, so it's a real mystery.

The f_C represents the fraction of civilisations that develop means to communicate or release signals into space, which could be detected or intercepted. This is the fraction of alien civilisations that develop the technological capability to start both the SETI (Search for Extra-Terrestrial Intelligence) and METI (Messaging Extra-Terrestrial Intelligence)—the ability to send and receive communications across *inter-*

stellar space (from the scientific terms *inter-* for 'between' and *stellar* for 'star', meaning 'the space between stars/planetary systems').

There is no way of knowing what this fraction could be, whether this sort of behaviour can be expected—if curiosity, communication and the search for others are qualities intrinsic to intelligent life, or if they're uniquely human. The number that Drake originally proposed was 0.1–0.2 (that is, 10–20%) of intelligent civilisations would develop these sorts of technologies; these estimates are still used today.

L

Finally, *L* represents the length of time (in years) that a sufficiently advanced civilisation with the technology to communicate is *actually communicating*. How long do we have (in real-time) to send and receive signals with this civilisation? Or, to say it another way, for how long can we detect the communications being emitted from this planet? For life on Earth, we've only been using radio waves since the late 19th century, and these weren't targeted into space. Even so, if we're generous and count them, *L* is only about 100 years here.

That's the Drake equation in all its glory. It's fairly involved, and there's a lot to go through, but each element makes sense individually and when taken in concert, the result gives us a rudimentary means by which to quantify the number of intelligent communicative life forms in the Milky Way. We just need to work out or estimate some reasonable values for all these variables.

Now, I can't comprehensively describe the Drake equation and its various intricacies without pulling some hypothetical numbers together to get an estimate; to quantify how many and how common other extra-terrestrial intelligent civilisations are. Let's do it now. We'll use some of the more accepted numbers for each variable to get a value for *N*.

VARIABLES	RATIONALISATION	VALUE
R_*	We've determined this to be about 3 stars per year.	3
f_p	We're starting to understand the planetary formation process in greater detail and with more confidence. Coupling this with the larger size of our observation sample (more than 10,000 potential planets and more than 3000 planetary systems), we can see it's atypical for a star not to have any planets. As such, we expect this number to be near enough to 1 that we approximate it as 1; that is, all stars (except in unique, exotic cases) harbour a planet.	1
n_e	The existing research and observations indicate that a value of 20% is reasonable and well accepted.	0.2
f_l	We're not exactly sure what fraction of Earth-like planets go on to develop life, but the leading hypothesis still leans heavily towards it being an inevitability; that is, $f_l = 1$. Thus, the current consensus is that life naturally develops when the conditions are right, like what you would find on an Earth-like planet.	1
f_i	What fraction of planets that develop life go on to become intelligent? Where do you even start with a number like that? Well, there are two options: (a) A planet has all the ingredients for life; eventually, it evolves life and then, on a long-enough timeline, that life evolves to become intelligent. (b) A planet has all the ingredients for life; eventually, it evolves life and then, that life evolves but not necessarily towards intelligence. I think evolution is often misunderstood, with this underlying thought that intelligence is the end goal. But life existed for hundreds of millions of years without getting smarter—just look at the reign of the dinosaurs. Another side of the same coin is that natural selection doesn't have an agenda.	0.5

	Life doesn't want to get smarter; it just wants to continue to exist, usually by getting better and more optimised. It just so happens that the mutation that led to our increased intelligence and cognition yielded alarmingly dominant results. The fact that we've dominated every other species to become the rulers and custodians of Earth isn't because nature knew that intelligence would have this result. It was just a random permutation that yielded it. When it comes to evolution, intelligence isn't necessarily an inevitability. What does it all mean? Well, without any strong scientific work reasoning either way, we just have to split the difference at 0.5: there's a greater than 0% and less than 100% chance that life evolves intelligence (I mean, it should be less than 100%—the dinosaurs had several hundred million years to evolve intelligence and made little progress), so a halfway point here is reasonable.	
f_c	f_c is basically the ratio of how likely we think a civilisation will develop the abilities to be communicative. It's difficult to assess this number quantitatively, but let's make the assumption that *'curiosity'* and *'the search for others'* are intrinsic ideas to intelligent civilisations. Life, at a sufficient level of intelligence, wants to communicate and see if there are others out there. They want to gain a sense of belonging and understanding. As such, we'll assume that 10% of civilisations will develop the ability to communicate beyond their planet.	0.1
L	How long will a communicative civilisation be spewing out communications? We have one data point, which is the roughly 100 years we've been using radio waves (if you only count messages being beamed directly into space, it's even less). There's no easy number here. Scientists have said it could be thousands to hundreds of millions of years. Let's be optimistic. Let's imagine civilisations are able to kick around for 100,000,000 years (10^8 in index notation).	10^8

Let's now plug all these values into the Drake equation. What does it spit out?

$$N = R_* \times f_p \times n_e \times f_l \times f_i \times f_c \times L$$

$$= 3{,}000{,}000$$

This estimates there are 3 million Earth-like, well-developed communicative planets in the Milky Way galaxy. Okay, that's a pretty big number! Contextually, in a galaxy like the Milky Way with 100–400 billion stars, that 3 million isn't looking quite as big as we thought. It's about 1 for every 83,000 stars (or approximately 0.0012%). While 3 million seems like a lot, in the context of the Milky Way, it looks like the numbers are more stacked against us than we might have first thought.

By the Drake equation and some fairly optimistic numbers, it still seems as though advanced civilisations with communicative technology are quite rare: 1 in every 83,000 stars is not the best odds. Still, it's not zero! So, how do we go about looking for these places? Perhaps, we need to think more about what specific technology we need to harness to gain a better grasp of what's going on.

In Chapter 13, we'll look at the upcoming space missions that seek to explore those places in the Solar System that are most interesting from a perspective of *could life exist there?* We'll also find out how we intend to look for evidence of life in far-off worlds (exoplanets). And we'll have a bit of fun trying to predict (statistically speaking) what the aliens and their home planets look like.

CHAPTER 13

THE SEARCH

Going by the Drake equation in Chapter 12, we have estimated just how many abundant intelligent, communicative civilisations there should be throughout our Milky Way galaxy. Yet, here we are, sitting alone atop our shiny blue marble, pondering: Are we alone? What is the next step, then, in searching for life, aliens and interstellar civilisations beyond Earth? Well, it's not easy. Like in all science, we do things in a series of small steps, with each step gradually and iteratively building on our successes, or using our inconsistent results to help nudge us down a different path to continue to improve, refine and pull together a more robust and rigorous hypothesis.

We've been taking these small steps in Parts 1 and 2, setting the scene for what we're looking for (what makes Earth special), and how we're looking for it (where else is special). Now, we take the next step. Where are the aliens? How do we spot them? And what have we found?

Where, and how, would we start such a huge undertaking? We've got a bunch of space exploration missions on the horizon, such as the next generation of space telescopes for looking at exoplanets with higher fidelity and in more detail, and also advanced spacecraft to explore some of those fascinating icy moon worlds we discussed in Chapter 9. We've even got an ambitious project that aims to send laser-accelerated spaceships to other stars! Clearly, we're working on building the technology to look (and we'll discuss that soon enough), but what are we *looking for?*

What sorts of things are we looking for that would be indicative of life? Or indicative of an extra-terrestrial civilisation? And while we have all these powerful engineering wonders being developed to search deeper and further, can we—similarly to how we did with the Drake equation—use mathematics to make some predictions about what sort of world we're looking for? Or what sort of organisms could be living there? Could we make some purely mathematical guesses?

With the technology we've got on the horizon, what are the space missions we have coming up that aim to help us find places where life could exist, or find life itself? Let's first look at the missions that are either in progress, or being planned, which target the places in our Solar System we discussed in Part 2: Mars, Enceladus and Europa.

MISSIONS TO MARS

Let's start with Mars. The good news is, we've started to get pretty good at launching things to Mars. We've successfully landed several rovers on Mars and, as of April 2021, a little autonomous helicopter. Because it's close and we've had all this practice, we've been launching bigger and more sophisticated rovers to Mars—the latest, *Perseverance*, is the size of a car. It's an engineering marvel. The exciting part about *Perseverance*—or I should say most exciting part (it's all exciting)—is that its core goal is to investigate whether Mars was habitable in the past and whether we can find any evidence or signs of ancient life. I would love nothing more than that, by the time you are feasting your eyes on this book, the first science is starting to filter back from *Perseverance* with evidence of some of the things we've discussed in this book. Evidence of ancient, fossilised life and a habitable world from the past.

We can't get ahead of ourselves, however; this may not have happened yet. As we've discussed, not finding something can be just as valuable as finding something. We still get information and that helps to paint a better picture of the world around us.

EXPLORING EUROPA

On the red planet, we're actively searching for signs of habitability or life in days past. What about the two moons we looked at, Enceladus and Europa? Let's look at Jupiter's Europa first, because it's the next in logical order by distance from Earth. We've ascertained that Europa has a subsurface ocean, and we could get some amazing insights if we could take a closer look at the icy marble. Fortunately, NASA has greenlit a new mission, the *Europa Clipper*, to do exactly that. The *Europa Clipper* is a space probe that will be sent into orbit around Europa. While in orbit, it will use its vast array of onboard instruments to take measurements and imagery to help us gain a better understanding of Europa's geological features, as well as understand the nature and chemistry of its subsurface ocean. Remember the *Mars Express Orbiter* we discussed in Chapter 8? As it orbited Mars it was able to explore below the red planet's surface all the way from space using sophisticated instruments. *Europa Clipper* would ideally be able to deploy a similar approach to unveil the mysteries below Europa's icy surface from its vantage point in space orbiting the icy moon.

The *Clipper* has a planned launch for 2024; hopefully, it will answer some of the questions we've put forth in this book about Europa's subsurface ocean. Understanding Europa's chemistry more could put it in a similar position to Enceladus: another world that looks to have all the ingredients for life to exist. We also mustn't discount how impactful mapping the geological features of Europa would be. Pending the success of the *Europa Clipper*, a follow-up mission called the *Europa Lander* aims to land a probe on the surface of Europa. At the moment, however, this is a bridge too far—we just don't have enough understanding of that moon's surface, so it could be a tremendous waste of resources if we try to land with the wrong equipment. That's why NASA is taking a two-stage approach: the *Clipper* will first get a more detailed look at Europa from a safe distance, then the *Lander* will make contact with its surface and get an unprecedented front-row seat to Europa itself.

ENCELADUS AWAITS

Next is Saturn's Enceladus. Unfortunately, Enceladus doesn't have its own mission yet. But there are several proposals, two of which aim to investigate Enceladus from an astrobiological perspective: the Enceladus Life Finder (ELF) and Enceladus Life Signatures and Habitability (ELSAH) missions. Both ELF and ELSAH would study the chemistry detected in the water volcanoes erupting from Enceladus's surface, so we can better understand the origin of these molecules, and learn more about the internal structure of its subsurface oceans.

It can't be overstated what a big undertaking these missions are. Jupiter and Saturn *aren't close*. These missions would take years to reach their targets, so it's important that we get them right. So, while it can be frustrating that these missions are few and far between, and often have limited scope, it's because of the challenges that must be overcome in putting spacecraft in orbit around those moons (let alone land on one of them!). It's hard work.

Despite that, of the three places within our own Solar System that we've discussed, we're actively conducting observations on one (*Perseverance* on Mars), we're planning towards another (the *Clipper* on Europa) and are in the early stages of working out how to tackle the third (ELF and ELSAH on Enceladus). So if you were reading Part 2 thinking, *Oh, I wish we had the answers*, then you're in luck. *We're looking at these places right now.*

This isn't the stuff of science fiction anymore; we're observing and engineering solutions to answer the very questions we're talking about in this book.

BILLIONAIRE-BACKED BREAKTHROUGHS

Something that is a little more science fiction is being developed in the private sector. The so-called *Breakthrough Starshot* project is a billionaire-backed initiative to send un-crewed spacecraft to our nearest star system, Alpha Centauri. The initiative proposes launching tiny nano-spacecrafts with lightsails into space (lightsails are exactly what they sound like: sails being 'pushed' by light, which are analogous to the sails of a boat being pushed by wind).

A ground-based array of ultra-high-powered lasers would shoot at the lightsails and accelerate these tiny crafts to a fraction of the speed of light (one-fifth the speed of light, in this instance). At one-fifth the speed of light, the travel time to Alpha Centauri is cut down to only 20 years. This means that, within our lifetime, *Breakthrough Starshot* could capture the first-ever images of another star system!

If you're sitting in stunned silence, mouth agape, that's about par for the course when first hearing about the objectives of this audacious mission. It is such an aspirational goal, which I hope is realised in my lifetime but, at present, it genuinely still sits in the world of science fiction. Several technologies require a significant step-up in capability (we're talking orders of magnitude), but they're being worked on. It sounds like something out of *Star Wars*—laser-propelled interstellar spaceships!—but I love that it's getting real attention, real funding and, most importantly, real research to try and make it happen.

A LONG-AWAITED EYE TO THE UNIVERSE

The next space mission has finally become a reality and is much closer to home: it's the successor to the Hubble Space Telescope, The James Webb Space Telescope (JWST). In astronomy circles, the timing of the launch of JWST was a bit of a sore point. It's been plagued by delays for years, but it just so happens that the JWST successfully launched in December 2021, fully deployed its mirrors in January 2022, and is currently aligning and calibrating its onboard instruments (as of March 2022).

JWST will be able to achieve the same things that the Hubble Space Telescope currently does, *just better*. It has bigger and better instruments across the board (and rightly so, because Hubble launched more than 30 years ago!) and should give us deeper and richer insights than we've ever had before. JWST will be able to observe exoplanets with unprecedented detail. This doesn't mean photos (sorry to get your hopes up)—these exoplanets are just too small, too faint and too far away. The detail will come from the various types of observations we've discussed earlier. In particular, using the transit method to learn about the atmosphere of planets. JWST will let us understand the atmosphere of small rocky exoplanets with a level of detail we've never seen before. This means we will continue to learn just how Earth-like (or un-Earth-like) some exoplanets really are.

SIGNS OF LIFE

To really understand the power of a telescope like JWST, we need to discuss what exactly we're looking for in a planet's atmosphere. We're looking for something called a *biosignature* and, if we break up the word into its constituent parts, what that is quickly becomes clear: bio- as in 'biological', and signature as in 'a unique signal or pattern'. So, a biosignature is a unique pattern or signal of biological origin. To put it another way, biosignatures are chemicals that indicate the presence of life (past or present) because they don't occur spontaneously in nature without life creating them. We've mentioned a few times how nature has preferences; it likes to do things a certain way. In Chapter 7, we saw how life carries out reactions that require energy (the example of trying to inflate a balloon that's already inflated). What this means is that, sometimes, life can perform chemical reactions that nature wouldn't otherwise perform. Life can create chemical by-products that nature wouldn't create otherwise, so their presence in an atmosphere would be evidence of the existence of life. Just to clarify, when I say 'chemical by-products of biological origin'—yes, I really mean we're looking for alien farts.

Let's think about this from the perspective of Earth and oxygen. Oxygen is in reasonable abundance in the atmosphere (making up about 21%), but it hasn't always been that way. Oxygen as it exists in the atmosphere, O_2 (that is, two oxygens bonded together—we discussed bonding in Chapter 1) tends not to exist in great abundance, because it is highly reactive with metals. You'll have seen this phenomenon with your eyes already: rusting. Rusting, or *oxidising* scientifically speaking, is the process where oxygen bonds to metals to form oxides. Oxygen *loves* to do this, so when it exists in the atmosphere, it rapidly bonds to a metal as a solid, but then it's no longer floating about for us to breathe.

Before panic sets in, *this doesn't mean we're running out of oxygen!* This is because of the exact thing we've been discussing: biosignatures. Life can create chemical by-products that nature wouldn't create otherwise (or in abundances beyond what nature wants). Nature wants oxygen to oxidise and form rust, but there are also some critical life forms that keep spewing out oxygen faster than rust can form. Plants, algae, plankton and certain bacteria all live their lives to the fullest by pumping out loads and loads of oxygen. They do this so quickly, and in such large abundances, the rust can't keep up.

While oxygen is getting ripped out of the atmosphere to oxidise metals—and we know this because we see things rusting with our own eyes—nature does so only at a particular rate, which will change depending on how much oxygen is actually in the atmosphere. At some point, a natural equilibrium is found, where the rate at which oxygen is being belched into the atmosphere by these organisms matches the rate that oxygen is getting extracted to rust metals. Under present-day conditions on Earth, this equilibrium happens when oxygen makes up about 21% of the chemicals in our atmosphere, and so that is where it sits: nice and stable at 21%.

But this wasn't always the case; the amount of oxygen in the atmosphere has changed considerably over the last few billion years. Typically, however, the amount in the atmosphere relates to the abundance of life on the planet (and vice versa).

Does that make oxygen a biosignature? The answer is ambiguous: *sort of*. It's true that an abundance of oxygen in the atmosphere would indicate it's being pumped out faster than it's oxidising, and that the means to pump it out would be of biological origin. The problem arises from how to detect oxygen (O_2) without mistaking it for a bunch of other molecules that just happen to have oxygen in them. And therein lies the rub.

A biosignature must be a chemical of biological origin, but for it to be useful to us it *also* has to fall within our detection limits. The oxygen that exists in our atmosphere is easily mistaken for other oxygen molecules, such as carbon monoxide and water, neither of which are of *only* biological origin. So, while we know what we're looking for, in terms of what we need the chemical to represent, it's remarkably difficult to pinpoint a chemical that does all this and can be detected reliably without mistaking it for something else. Oxygen (O_2) and ozone (O_3) are still considered high priorities as biosignatures, as well as nitrous oxide (N_2O) and methyl chloride (CH_3Cl), but this is by no means an exhaustive list and, as mentioned, they are still challenging to detect reliably.

WHAT COULD LIFE LOOK LIKE?

We've looked at some of the upcoming space exploration missions that target our own Solar System (*Perseverance* for Mars, the *Europa Clipper* and *Lander* for Europa, and ELF and ELSAH for Enceladus) as well as other star systems (the JWST and the straight-out-of-science-fiction *Breakthrough Starshot*). After these missions, of course, there will be more. With each generation, the technology improves, and grants us a deeper and richer look at the worlds beyond Earth. Now, wouldn't it be cool to imagine what the life on one of these far-off worlds would look like? In our own Solar System, where we have many high-resolution observations available, we expect that *if* life existed (for example, ancient life on Mars) or if it *still* exists (for example, in the subsurface oceans of Enceladus or Europa), it would take a very simple form. Microbial, maybe unicellular.

When looking beyond at worlds that are dozens of light-years away, however, we don't necessarily make the same assumptions. Because we can't see in nearly enough detail (we're still looking for signs of life, let alone details about the level of complexity of that life), theoretically, the life could be much more complex. It could be multicellular, land dwelling or intelligent. The probability (as we discussed in Chapter 12) is unlikely, but let's pop on our creative hats for a bit. We'll keep a foot in science—we're not just going to make things up or wildly speculate—but we'll have some fun with the *what ifs!*

Let's start right now using inspiration from a published scientific article to try and predict using statistics what aliens might look like, and what sort of planet they might live on.[5]

It may seem like a fruitless endeavour to try and make predictions from just one data point (us), but this is where statistics become a powerful ally. We can learn quite a bit about how we fit into the Universe by considering that we have an increased likelihood of belonging to certain groups, and how those groups relate to what might be considered 'normal' or 'average'. This still all sounds vague, so let's use an analogy to help get our heads around the idea.

Consider picking a person randomly on Earth. Which country are they most likely from? Due to the larger population size, you'd be correct in saying it's most likely they'll be from China because it has a population of about 1.4 billion people. This accounts for about 18% of the global population, or in other words, there is roughly a one in six chance that a randomly selected person on Earth is from China. In fact, more than a third of the world's population is concentrated in countries with more than 1 billion people (China and India). Check out Figure 13.1.

Here's the thing, though. While the probabilities suggest that a randomly selected *person* is most likely to be from China, when compared with every country on Earth regarding population, China is not what you would consider 'typical'. There is only one other country in the world with a population above 1 billion, India, and there are about 200 countries in the world, so countries with more than 1 billion people only make up about 1% of all countries. Something that only makes up 1% of the entire sample is clearly more of an outlier rather than the 'average' or 'typical'. So, if you randomly selected a *country* out of the list of some 200 countries, you have quite a small chance (1%) of selecting China. You'll see this in Figure 13.2.

We can take this a step further by considering what the median or 'typical' country population size is. The median country is the one which, if you lined up all the countries in order of population size, would be in the middle. Turns out it's countries with a population of around 6 million, so for example, Denmark.

The interesting thing about how the world's population is distributed is that even though half of the countries are larger than Denmark and half are smaller, 98% of the population is in the larger half. Take a look at Figure 13.3. So even though you have a 50% chance of picking a country smaller (by population) than Denmark if you selected a country at random (remember, Denmark is smack bang in the middle, it's our median or 'typical' country by population), you only have a 2% chance of picking a *person* in a country smaller than Denmark if you selected a person at random. So, there is a rather strong selection bias towards people belonging to high-population countries.

What does this mean? It means that, if you randomly selected a country, you have a 50:50 chance of selecting a country greater than or lesser than 6 million. Figure 13.2 shows that, nearly two-thirds of the time (63%), we'll actually end up picking a country with a population between 1 million and 100 million.

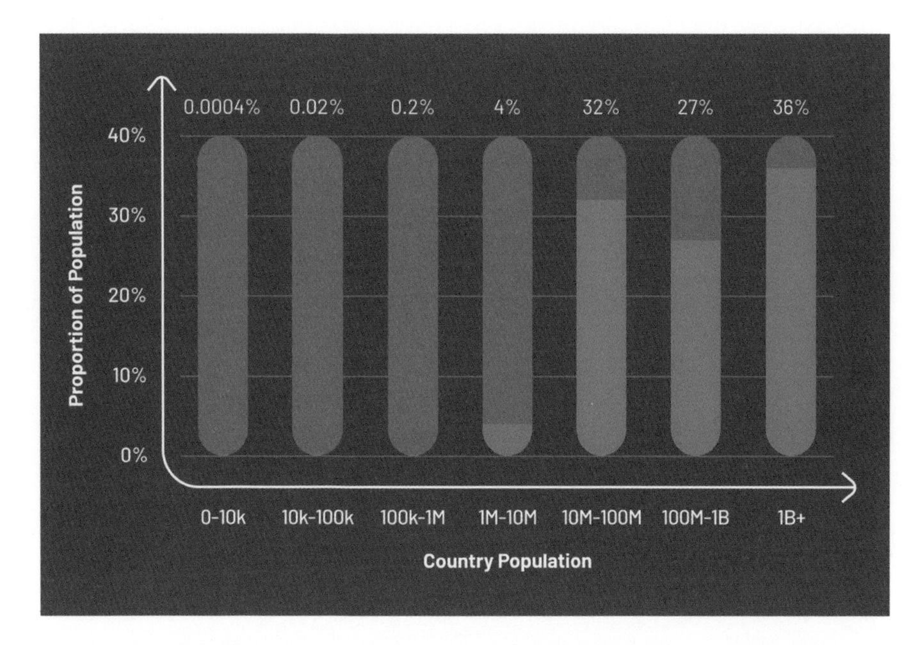

Figure 13.1: The distribution of the global population between countries of different populations. Over one-third of the world's population is concentrated in countries that have a population of more than 1 billion (China and India).

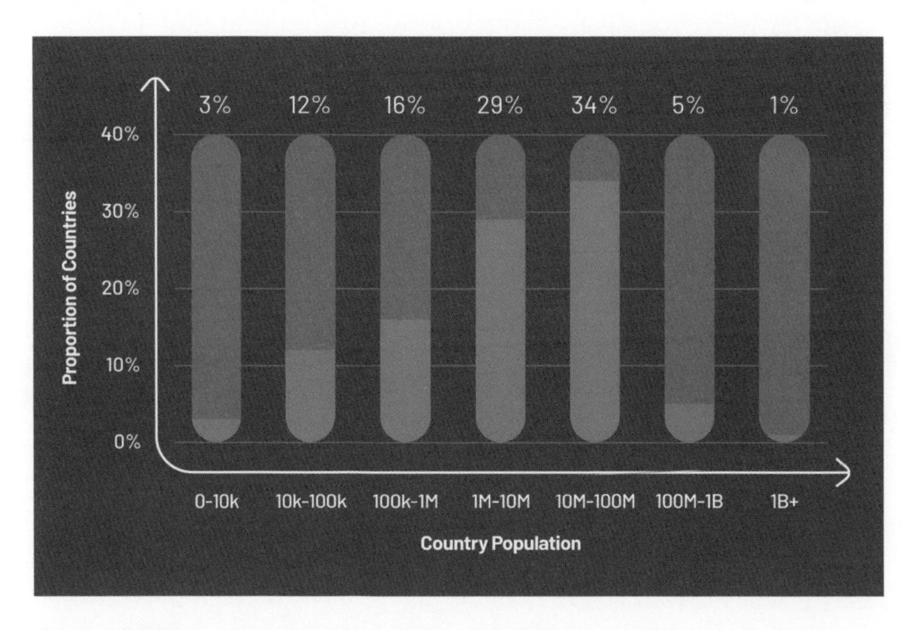

Figure 13.2: The proportion of countries with different populations. Countries with more than 1 billion people make up only 1% of all countries; most have populations of either 1 million–10 million (29%), or 10 million–100 million (34%).

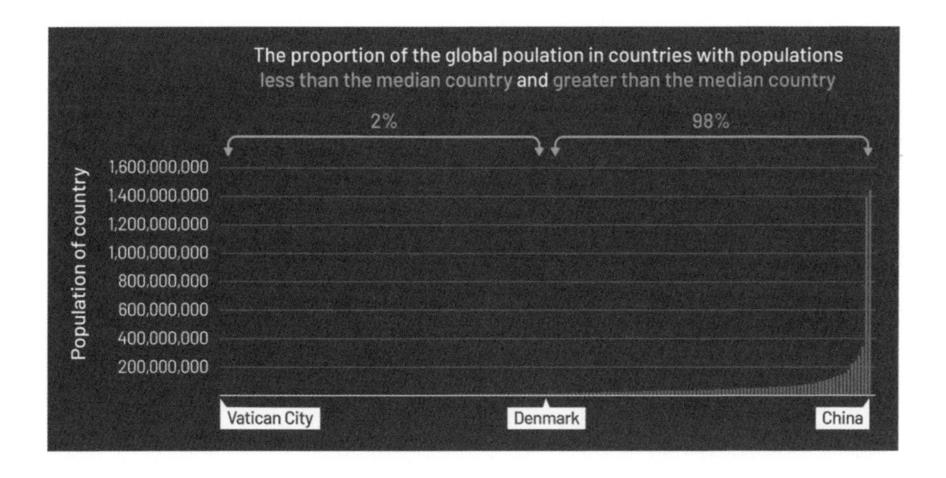

Figure 13.3: All the countries lined up in order of population, with Vatican City having the smallest population, Denmark in the middle, and China having the largest population. You can see that even though Denmark is in the middle, larger countries make up 98% of the global population.

How does this all relate to searching for aliens? Well, it can be a bit tricky to wrap your head around, so try considering the following step by step.

1 A person is more likely to be born with a common hair colour than a rare one.

Fair enough, right? Hair colour, eye colour, blood type, whatever the property—you're more likely to have the more common property than the rare one. You're more likely to be born right-handed. You're more likely to be born with brown eyes. It all makes sense so far. Let's extend this to place of birth, then.

2 A person is more likely to be born in a higher-population country than a lower-population country.

Not quite as intuitive as step 1, but we've shown earlier that there's a bias towards people belonging to high-population countries. If 'country of birth' was a property, then it follows the same logic that you're more likely to be born with the more common property (being from a country with a larger population than Denmark) than the rarer one (being from a country with a smaller population than Denmark). Let's take the next big leap.

3 A person is more likely to be born on a higher-population planet than a lower-population planet.

Again, not as immediately intuitive but the premise is the same. It's much more likely you are born on a higher-population planet than a lower-population planet, in the exact same way as you are more likely to be born in a higher-population country.

So, what does this all mean? At its simplest, it means that Earth is more likely to be a more highly populated planet than what is the 'median' or 'typical' planet. Earth is more like China, India or the United States—places with a hundred million to a billion strong populations. It suggests that we're more likely to be searching for the planet equivalent of Denmark (the median or 'typical' country with regard to population size)—a planet that falls in one of those big columns in Figure 13.2.

By using some simple statistics, we've shown that there is a higher likelihood that other worlds have populations smaller than our own. Is this a certainty? Absolutely not. We're trying to extrapolate information from a single data point. In the same way that the probabilities suggest a random person selected on Earth has a high likelihood (98%) of living in a country with a higher population than Denmark, it's random and so they could still end up being from a country with a population smaller than Denmark (even if the likelihood is only 2%). So, we're certainly not definitively saying other planets with life have fewer inhabitants, but that it would make a lot of sense probabilistically speaking. Probability can still paint a picture of what other worlds might look like.

That's really cool though, right? Using statistics, we can predict what sort of population an alien civilisation might have on their home world!

THE BIRTHDAY PARADOX

Statistics and probability are powerful tools across a wide breadth of applications, from medical science to meteorology. Being able to determine the likelihood of an event occurring is almost like having a crystal ball to see into the future. Whole industries are built around being able to understand how risky certain actions, behaviours and circumstances are.

They also have really fun applications. One of my favourites is what's referred to as the Birthday Paradox because it seems so unintuitive but, once you understand it, you can really boggle your friends' minds with it!

The basic premise revolves around the following question: In a group of x people, what is the probability of two people sharing a birthday (not including the year)? The knee-jerk reaction is, well, obviously x/365, right? So, if there were ten people in a group, it would be 10/365 or about 2.7%. A low likelihood. Unfortunately, this is wrong. The reason is because we have a tendency to think only about the odds of someone having the same birthday as ourselves. The reality is that it's not just you and person 1, you and person 2, you and person 3 . . . all the way up to you and person 10, who are the pairs for comparing birthdays. It's also: person 2 and person 3, person 2 and person 4, person 5 and person 8, person 9 and person 3, person 7 and person 8 . . . in every permutation. You see how many more pairs there are? The number of all possible pairs grows exponentially.

If you do all the maths, it turns out that, in a group of 23 people, the odds of two of them having the same birthday (sans year) is just over 50%. That seems so bizarre, but the maths checks out. And it only goes up from there. In a group of 30, it's about a 70% likelihood. In a group of 50, it's 97%!

If you want to have some fun with your friends, the next time you're in a group of 23 or more people, you can show off your predictive abilities. You've got a loaded die, but the others won't know that!

We can take this a step further, because the population size can tell us more information about the planet and, most interestingly, the aliens themselves. There's a correlation between population size and country area. Typically, the larger the country, the greater the population. Obviously, there are exceptions to this, but there's a strong enough correlation to be useful for our purposes. We should expect that alien life would populate a planet in the same way; that is, the size of their population grows to reach an equilibrium with the planet. Therefore, from our previous prediction that the alien world will have a smaller population compared to Earth, we can also hazard an estimated guess that the planet is also likely to be *smaller* than Earth.

In a similar fashion, there's a correlation between organism size and population. Typically, the smaller the animal, the more abundant the population. Ant species boast populations in the billions, whereas there are only a few hundred thousand African elephants, and about ten thousand blue whales. The huge amounts of energy that large animals require dictate that their populations can't increase dramatically (they physically take up more space, too!). In this way, if the alien population is smaller than ours, we can infer that they're also likely to be *bigger* than we are.

That's pretty wild, right? From just the one data point (humans existing on Earth), we can use statistics to try and guess that alien life is most likely to be:

1. bigger than we are
2. living on a smaller planet than Earth
3. having a much smaller global population.

All that just because of some pretty neat little tricks with statistics and the likelihood of us belonging to a larger population group than the 'typical' planet.

We ploughed through a lot of stuff in this chapter, and we've now started to discuss (while remaining firmly planted in science) some of the captivating areas of exoplanetary exploration that are closing the gap between reality and science fiction. While we're on a roll, we'll continue in a similar vein. In Chapter 14, we'll consider some interesting things we've observed, which may be of alien origin; we'll also dig more deeply into the idea of advanced alien civilisations.

CHAPTER 14

EVIDENCE

We humans are in awe of the cosmos; it's likely we have been since we've had the ability to think. The diamond-studded black canvas sparkling above that is the night sky is a beautiful sight, and the darker our surroundings the brighter and more dazzling a show the night sky puts on for us. It's no surprise that it has been a fundamental part of the human experience and has often been tightly intertwined with spirituality and rich stories about our origins and our place in the world.

As we've learned more about ourselves, the world around us and the stars beyond in this book, we've discovered that these stars represent other planetary systems, many (if not all) of which harbour planets, some of which may harbour life, and even fewer still may harbour intelligent life capable of communications or space exploration. In the previous chapters, we discussed the likelihood of some of these things occurring, but also what efforts are either in place or on the horizon that aim to explore some of these worlds in sufficient detail to inform whether or not life may exist. If we're taking these steps to explore other planetary systems in our neighbourhood, others must be doing the same. Have any intelligent aliens tried to contact us? Or have their advanced engineering feats been visible to us in other forms?

Let's have a look at some of the interesting things we've seen, and how they could be considered to be of alien origin, or if there are other logical explanations.

THE ARECIBO MESSAGE

As we touched on briefly in Chapter 12, we've been spewing radio signals into space for quite some time. We sent the first 'message' into space in 1974, the so-called *Arecibo message*, aptly named after the Arecibo radio telescope broadcasting it. The message itself was less than a quarter of a kilobyte of data, and it was cleverly encoded with details about our numbering system, the chemicals that make up our DNA, details about us physically and our location within the Solar System. It's all encoded in binary; that is, 1s and 0s, and would likely be incredibly challenging to decipher, especially if you consider that decoding it requires an understanding of the binary number system. Regardless, it was sent out almost half a century ago, and because radio waves are a form of electromagnetic radiation and thus travel at the speed of light, it will have reached almost 50 light-years away from Earth.

This wasn't the first signal we beamed out, however. Inadvertently, every radio signal we've been sending around the world will also have been transmitted off the planet. Radio waves are broadcast omnidirectionally to reach a wide audience, which also means they're blasted upwards. Granted, these signals will be significantly weaker than one-directional communication, but they are still being sent into space. By that definition, we've been sending radio waves into space since just before the turn of the 20th century. That means we've sent signals into space around Earth approximately 120 light-years in any direction. At that distance, the signals would be incredibly weak, so whether a nearby intelligent civilisation could detect them would be limited by the power and sensitivity of their detection technology.

Back to the Arecibo message, though. We've highlighted that the contents of the message would be puzzling, bordering on arcane, if a civilisation received it, and they may never understand what the message says. But there's something in the message that they will understand, without a doubt, and it doesn't depend on them understanding binary or the message's particular format. *The signal is artificial*. By artificial, I mean that it has been manufactured; a signal encoded like this isn't something that would occur because of any natural event. Signals that occur naturally are smoother; they have natural oscillations and frequencies that nature dictates; and they manifest with predictable properties and traits. The Arecibo message has been designed. It is artificial. An advanced civilisation would recognise that there is something unnatural about this message and so, while they may not be able to translate its 1s and 0s into something intelligible, they could appreciate that it is likely the product of another civilisation.

TECHNOSIGNATURES

This sort of artificial signal is called a *technosignature*. Just like its biosignature counterpart, a technosignature is a signal indicating the presence of life but, instead of doing so through biological chemicals, it uses technology. To other civilisations, the Arecibo message would be considered a technosignature. It has been produced by technological means, not as a by-product of a naturally occurring event. Along with signals, a technosignature can take the form of different engineered objects such as structures, spacecraft or artificial light. So, if we've beamed out our own technosignature to be detected (as well as the faint signals we've been broadcasting for more than 100 years), that begs the question, have *we ourselves* observed any technosignatures?

Back in 1977, we *did* detect something. A radio telescope called the *Big Ear* detected a strong signal that lasted for 72 seconds, coming from the direction of the Sagittarius constellation. Astronomer Jerry Ehman was so blown away by the signal (because it exhibited the characteristics of a technosignature from another civilisation), he wrote 'Wow!' next to the signal printout, thus inadvertently anointing it the *Wow! Signal.* The Wow! Signal was powerful, more than 30 standard deviations above the background noise (this is called the signal-to-noise ratio).

When it comes to detecting things in astronomy, this is considered a *very* strong signal (typical observations require a signal-to-noise ratio of 5 or more to be considered reliable, so 30 is incredibly strong). It didn't contain any information—it wasn't an encoded message like the Arecibo message we had beamed out—but it had a frequency that was eerily close to the natural frequency of hydrogen. Now, astronomers generally agree that an extra-terrestrial civilisation may well use the natural frequency of hydrogen to transmit radio signals. This is because it's such a fundamental frequency that advanced civilisations might ubiquitously realise it would be a way of signalling intelligence without using an encoded message, and also because it travels extremely well through space. Such a powerful signal at that specific frequency screams technosignature; that it is of extra-terrestrial origin. Since the Wow! Signal was detected, however, there has been nothing more. No further observations, despite rigorous searching and listening.

So, what happened? There have been several hypotheses, but none have categorically iced it. One hypothesis suggested it was hydrogen surrounding two comets, but this has since been disproven.

Perhaps the least exciting answer is that it was a signal originating from Earth that was reflected or bounced back by space debris (there is a *lot* of stuff orbiting Earth). The frequency at which it was detected is a restricted bandwidth to broadcast on, however, so there are problems with that hypothesis, too.

The reality is that, right now, we don't have a clear answer for the signal's origins. Perhaps it was an alien lighthouse that only completes a rotation of its 'light' once every 50 years, and the next Wow! Signal is only a few years off (as of writing this). As with most things we'll discuss here, it's fun to talk about the alien hypotheses but, in astronomy circles, the general answer when we spot something unknown is: It's not aliens. This is unlikely to be the exception. But it's never aliens until it is, right? So we'll keep exploring these fun hypotheses, accepting that the likelihood of it being alien in origin is astronomically miniscule (we'll go through the generally accepted scientific hypotheses, too).

ET TV (EXTRA-TERRESTRIAL TELEVISION)

Ever since humans first used technology in the form of crude tools, we've continued to make technological progress as we invent newer, better, faster and more advanced tools to do the job. Observing the skies is, of course, no exception. One of the grandest undertakings in the realm of observational astronomy is taking place right here in Australia, the aptly called *Square Kilometre Array (SKA)*. It's all in the name really—an array of radio telescopes that covers a square kilometre. The SKA will be in esteemed company; Australia has a decorated history in radio astronomy and boasts world-renowned facilities.

The SKA has several science goals and, while they're all important and profound in their own right, the most relevant to our discussion is the *Cradle of life: Searching for life and planets*.

Being so large, the SKA will be able to search for and detect a myriad of possibilities, including the chemistry that's critically important to life and extremely weak radio signals. Just how weak are we talking here? The SKA is so sensitive to radio signals, it will be able to detect signals on planets within a few dozen light-years, which have the equivalent strength of a television transmitter. They don't even have to be beaming the signal in our direction, the SKA is sensitive enough to detect the signal 'leaking' from the broadcast (like our omnidirectional radio broadcasts we discussed with reference to the Arecibo Message above). So, if one of our nearest exoplanet neighbours has life, *and that life is creating and watching TV*, the SKA will be able to detect it!

TABBY'S STAR

While we're discussing interesting observations people have thrown alien hypotheses at, here's one that takes the form of something we explored back in Chapter 10—a transit signal. If you recall, we use the transit method to detect exoplanets orbiting other stars. During its orbit, if the planet crosses our line of sight with the star, we will be able to observe a measurable 'dip' in the brightness of the star because some of its light is being blocked.

A paper published in 2015 showed that the Kepler Space Telescope (one of the heroes of space telescopes as far as exoplanet detections go—it's detected well over 2500 exoplanets!) had observed a dip in brightness around star KIC 8462852. Fortunately, the star also goes by an easier-to-remember name, *Tabby's star,* after the astronomer who observed the transit signal, Tabetha Boyajian. The observations took place over several years and there were two oddities that made the changes in brightness around Tabby's star so curious.

The first observation was that the dips were irregular. If you remember from Chapter 10, a planet orbits its star more or less like clockwork. We orbit the Sun every 365 days, Mars orbits every 687 days, Jupiter every 11.86 years and, when an exoplanet orbits its star, it will typically do so every x number of days. This will correspond with the dips in the transit signal with the same cadence. The problem with Tabby's transit signal was it wasn't regular like clockwork; it was erratic and irregular. That is weird.

Once we take it in concert with the second odd observation, we get a deeper understanding of just *how* weird it was. The second oddity is that these dips weren't small. They were big. *Really* big. In Chapter 11, we mentioned that a planet the size of Jupiter would block out approximately 1% of the light of a star. Well, these dips were an order of magnitude larger. Whatever was blocking out the light from Tabby's star was blocking up to 20% of its light. This thing was *big!* Something as big as Jupiter only blocks out 1%, and this was stopping 20 times that? What could do that?

That's the real conundrum of Tabby's star: *What is blocking out such huge swathes of light?* Over the years, there have been several hypotheses to explain the atypical observation. First, of course, is the alien megastructure hypothesis. This idea is that what we've stumbled upon is evidence of an alien megastructure (or potentially the early stages of its construction) around Tabby's star. One such proposed megastructure is the so-called *Dyson sphere*, an enormous structure built around a star in a sphere, so as to harvest the entirety of the star's energy, perhaps by means of a giant spherical shell that encases the star and that is completely lined with solar panels. Figure 14.1 shows an artist's impression of a Dyson sphere being built around a star.

Figure 14.1: An artist's impression of a Dyson sphere being built around a star. The sphere would enclose the star and be able to harness its energy, enabling an advanced civilisation access to tremendous amounts of energy.

Unfortunately, for you alien hopefuls reading, this is tremendously unlikely, if not impossible, mostly because the light being blocked comprises only specific frequencies and colours. To put it another way: if it was some kind of physical structure that had been engineered, it would almost certainly be opaque and thus block out *all light*. Because only certain colours are blocked, it's likely to be something more akin to an atmosphere blocking the light, like we discussed in Chapter 11.

If you recall, when light shines through an exoplanet's atmosphere, only certain colours are scattered or blocked (in those circumstances, we can sort of 'reverse engineer' which chemicals are in the atmosphere by the missing colours; our absorption spectrum and dancing eggs approach). Something similar is hypothesised as the likely culprit for these enormous transit signals. The leading hypotheses suggest that the culprit is perhaps clouds of comets are breaking up as they orbit the star; a residual dust or debris ring is left over after the star and planet formation process; or some sort of dusty, gassy or cloudy structure that is the result of a natural physical process. Nothing yet fits all the observations perfectly, but the alien megastructure doesn't fit *at all*. Still, it's interesting that these sorts of ideas, many of which are birthed from science fiction, are slowly starting to approach reality.

So far, we've discussed two interesting signals we've detected that, while extraordinarily improbable, have had an alien hypothesis bandied about to explain them at some point. But has there been anything we've seen that isn't a signal? Something that might be a little more physical? Of course, I wouldn't mention this if I didn't have a delightfully interesting observation to share with you.

FIRST MESSENGER FROM AFAR

It all took place in 2017. Prior to this, everything we had observed within our Solar System had originated here. We've sent spacecraft to orbit around other planets; taken breath-taking images of the dwarf planet Pluto at the edge of Kuiper Belt; landed rovers on Mars, probes on Saturn's moon Titan, and even landed on a comet! We've explored a small fraction of the Solar System, but we've done so with curiosity and aplomb. Common to all these destinations that we've explored and observed is that they all originated from within the Solar System. They are all the result of the complex and enigmatic star and planet formation process.

We covered some of this process in Chapter 4, but we couldn't go too deep because it's such an enormous and complicated topic. The salient point is that everything we see in the Solar System is the result of this natural, somewhat chaotic, process. Even when we venture far out to Pluto, or when a comet comes hurtling into the inner Solar System to put on a brief dazzling display in the night, these objects are all children of the Solar System. So you can imagine our sheer excitement in 2017, when we detected something hurtling towards the Sun that originated from *outside* the Solar System!

The object detected in October 2017 was bestowed the name *'Oumuamua*, Hawaiian for 'first scout' or 'first messenger', to acknowledge the title it holds as being the first object to visit our little corner of the Milky Way from beyond. When we first detected it, 'Oumuamua was actually already heading out of the Solar System. It had serendipitously passed inside the orbit of Mercury and, subsequently, was sling-shotted around the Sun and launched on its merry way back out into the void of space.

We know this because we can 'trace back' the trajectory it took, using supercomputers based on its location and velocity when we first spotted it. Just like if you see a football flying through the air, your brain can quickly trace back the path it took, and figure out who originally kicked it (or at least, whereabouts they were when they did). What this meant, though, was that our window to observe the object was already closing. The further its journey took it from the Sun, the fainter it would become, and the harder it would be to observe and ascertain any information about the object.

Despite the small window for observing 'Oumuamua, we learned a great deal about it. One of the most striking observations was its shape: what many have described as 'cigar-shaped'. This is interesting because, well, we don't have any cigar-shaped objects in our Solar System. We've mentioned it but, at a certain size, objects tend to take a spherical or spheroid shape, and anything smaller is usually some oddly shaped rigid body. But to be *cigar-shaped* or (as was being entertained) *rocket-shaped* seemed too artificial to be natural. Could this object have originated from an extra-terrestrial civilisation? This hypothesis gained momentum when we determined that the object was experiencing some form of acceleration that was *not* due to gravity.

Ultimately, its acceleration was found to be what was expected from the gentle push of solar radiation (we can't cover all the physics of it, but the key point is that radiation from the Sun creates a gentle push). But this was snatched up as being 'by design'—as in 'Oumuamua was built as a lightsail, like those we discussed in Chapter 13 for Breakthrough Starshot. That alien engineers had designed 'Oumuamua to harness the force from a star to accelerate between systems.

Despite arguments to the contrary, the reality is that this is highly improbable. The shape is odd, to be sure, but for an object to be ejected from its own star system, it needs to be launched with sufficient speed to surpass the system's escape velocity (the velocity below which the object would remain 'trapped' in orbit around the star). It's therefore more likely that a violent collision caused 'Oumuamua to be ejected, which is also likely to yield shards or long, elongated cigar-shaped objects. Because the object is spinning the way it is (spinning head-over-toe along its 'long axis', not in a spiral like a dart), this adds credence to the fact that it's unlikely to have been 'launched' this way.

There is actually little, if any, evidence that indicates 'Oumuamua is artificial. Again, our imagination can often get the better of us and, while it's exciting to consider these more exotic explanations, we must maintain scientific rigour. The leading hypothesis is that 'Oumuamua is largely made of nitrogen ice and is, in fact, a fragment from a Pluto-like object—one of those icy objects in our Kuiper Belt—that has been ejected from a violent collision in another star system.

It's a massive discovery, and the science is important and impactful, so we mustn't discount its significance just because it *wasn't aliens*. We had front-row seats to observe an object from another Solar System up close. It has profound implications; arguably, the most notable is that this is clear evidence that objects, chemistry and physics in other star systems behave as they do in ours. A subtle yet powerful observation.

❋

Back to answering the original questions: *Have any intelligent aliens tried to contact us? Or have their advanced engineering feats been visible to us in other forms?* The short answer is no. As we've explored in this chapter, we've certainly seen some things we can't comprehensively explain. Some scientists (albeit a small minority) have explained these things by entertaining hypotheses that suggest extra-terrestrials were involved.

Within the astronomy community, however, the mantra for new observations that we can't fully explain is: *It's not aliens*. This is almost certainly true, not just in the scenarios we've discussed in this chapter, but the vast majority of all future unexplained phenomena. The truth is that there is insufficient evidence to suggest aliens were involved in producing any of these signals, or any of these proposed feats of engineering. As legendary astronomer Carl Sagan once said, 'Extraordinary claims require extraordinary evidence.' While it's fun to toy with alien explanations for these observations, the evidence must be *much* stronger than the rather flimsy cases put together here.

With that being said, if these unexplained events aren't evidence of aliens, then we've got nothing. Nada. Zilch. Not a crumpet. Absolutely zero evidence of any life beyond Earth. What does that mean? In the next chapter, we'll explore this question in more detail and look at some of the most intriguing answers as to why we haven't heard from or seen anyone else.

As physicist Enrico Fermi famously put it: '*Where is everybody?*'.

CHAPTER 15

FERMI'S PARADOX

Let's cast our minds back to Chapter 12 and the Drake equation. We went through the calculation to come up with an optimistic estimate of there being 3 million intelligent civilisations in the Milky Way with the ability to communicate beyond their planet. I know we said back then that it's not quite as high as we might've thought initially, especially in relation to how many stars there are in the Milky Way, but it's still a lot. When you sit down and think about it—the idea of 3 million intelligent, advanced, and communicatively active civilisations—it does feel as though the Milky Way should be absolutely teeming with activity.

Even if we rein the numbers back to some less ambitious values (say, we give civilisations a lifetime of only 10,000 years), we'd still have 300 active civilisations in the Milky Way. These civilisations could be thousands of years ahead of us in technological advancement. They wouldn't just be resting on their laurels; if they're advanced enough to communicate, surely, they would be exploring, too. Surely, they would be sending out probes and blasting electromagnetic signals and detectable communications all over the place. Surely, there should be a buzz of activity all around us, and yet, that doesn't seem to be case. There's nothing. Not a peep. Why not? What are we missing?

This conundrum comes up again and again among scientists; the most famous instance was a casual conversation between physicist Enrico Fermi and a group of fellow scientists. Fermi put forward the powerfully simple question: *Where is everybody?*

WHERE IS EVERYBODY?

At first, it can seem anticlimactic. Simple answer: We just haven't seen anyone yet. No big deal. But as you start to dissect things, the perplexity of the paradox starts to manifest. The numbers just don't make sense in the scheme of things. Let's go into more detail about what that means.

Consider how much technology has advanced in the last 100 years. If you took a smartphone back to 100 years ago, people's heads would explode at what you were showing them. It's a crazy amount of technological progress in such a short amount of time. Now, consider what that progress could lead to 100 years *from today*. Imagine what sort of technology could be developed 100 years in the future, which would have the same effect on you *right now?* What sort of technology would you just be so mystified by and in awe of, you'd stumble in disbelief? And that's only after 100 years of technological progress. Imagine 1000 years, 10,000 years, *1 million years!*

I bring up the incredible rate of technological development because, while the Earth is only 4.5 billion years old, our galaxy—the Milky Way—is 13.5 billion years old. We've just discussed how technology even 100 years from now will completely blow our minds, so imagine if a civilisation had a *billion*-year head start on us. Even a thousand-year head start would be enough to have technology that would appear to us to be borderline magic. The fact is, it doesn't matter if it were a 1000 years or 1 billion years, the amount of progress a civilisation could have made *before* we came into existence would easily be sufficient to achieve interstellar travel. So why haven't they popped by to say 'Hi'?

The logical argument against this is obvious. Maybe they have the same problems with achieving speeds at a fraction of the speed of light as we do. Maybe it just takes too long to reach other star systems, so they never reach us. The problem with this argument is again to do with the staggering numbers we're dealing with, in terms of the number of stars and amount of time having passed.

Let's do a little thought experiment. As we estimated in Chapter 12, the Milky Way has about 250 billion stars (I've split the difference between the 100 billion and 400 billion estimates). Let's imagine an advanced civilisation with the technological means to build an autonomous spacecraft that travels to a nearby star system, harvests resources (from planets, asteroid belts or otherwise), then builds two more copies of itself (itself being the autonomous spacecraft) that can go forth to explore two more star systems. I know this sounds like some fairly exotic technology, but consider that we went from a 12-second flight over 37 metres in 1903 (the Wright brother's first flight) to landing humans on the Moon in 1969. Yep, just 66 years! So, a sufficiently advanced civilisation with 1000 years on us could easily pull off an autonomous self-replicating spacecraft. But I digress, back to our thought experiment. Let's say our spacecraft takes 250 years to travel to the next star, and another 250 years to harvest resources, build and launch the two next generation of autonomous spacecrafts, for a total of 500 years. If these two spacecrafts were to then travel to two more star systems and continue the process, how long would it take for them to have visited every star in the Milky Way galaxy?

We can calculate this. After 500 years, we've explored two stars. After 1000 years, four stars. After 1500 years? Eight stars. Essentially, we're doubling every 500 years. At this rate, our autonomous spacecrafts will have visited all 250 billion stars in as little as 20,000 years! Such is the power of exponential growth. It is blisteringly fast. But unfortunately for this particular thought experiment, *too fast.*

This calculation doesn't take into account the size of the Milky Way (the distances the spacecrafts would have to travel), and the physical limitations we have on how fast the spacecrafts could travel, but we can cater to this constraint. All the stars in the Milky Way are concentrated within a disc about 200,000 light-years across, so we need to know how long it would take to cover this distance. For our estimate, let's assume the first 250-year journey was made by a spacecraft travelling to our nearest stellar neighbour, Alpha Centauri, which is 4.37 light-years away. Let's round that out to 5 light-years for brevity. So, 5 light-years in 250 years would suggest a spacecraft capable of travelling at 2% the speed of light. Considering that the ambitious laser-accelerated mission Breakthrough Starshot was aiming for 20% the speed of light, 2% seems a conservative estimate for an advanced civilisation. With an upper speed limit of 2% the speed of light, how long does it take to cross from one side of the Milky Way to the other? Quick maths tells us it's 10 million years. Because our spacecraft are only travelling for half the time though (remember: 250 years travelling, 250 years harvesting resources and building), let's double this number to 20 million to be safe. So, 20 million years. That's how many years our quick back-of-the-envelope calculations tell us it takes to send a spacecraft to every star system in the Milky Way (and colonise them in some capacity).

While this may seem like a long time, remember, Earth's been around for 4.5 billion years. It's merely a blip. Just think: 20 million years is only 0.4% of the lifetime of Earth. For context, if we squashed the Earth's lifetime of 4.5 billion years down into one 24-hour day, it's the equivalent of only 6 minutes. To colonise the entire galaxy!

Hundreds of advanced civilisations could have done this throughout the lifetime of the Earth many times over.

While this might seem like a fairly elaborate tangent, it's all pointing out that the distances don't really matter. Regardless of the limitations on how fast a spacecraft could travel, the reality is that there's been a ridiculous amount of time for an advanced civilisation to rise, explore and colonise the entire galaxy. With the even more ridiculous number of star systems leading to our Drake equation estimation of advanced civilisations numbering somewhere between the hundreds and the millions, the probabilities seem absolutely stacked in favour of intelligence arising elsewhere and spreading forth. So, where are they? Why haven't we seen them?

Even if they're long gone and became extinct for some unknown reason, there should be evidence of them somewhere. It is a bizarre, puzzling and somewhat frustrating conflict between the number of stars in the galaxy, the enormous amounts of time that life, intelligence and civilisations have had to evolve, and the dearth of absolutely anything. No evidence of advanced civilisations. No evidence of visitors. No aliens. No ancient artefacts. No errant communications. Nothing. Just silence.

LIFE BEYOND THE MILKY WAY

Just when you thought things couldn't get any more baffling, we haven't yet considered beyond the Milky Way. That's okay, though; let's do it now. How many galaxies are there in the observable Universe? It's hard to give specifics, but it's thought to be somewhere in the realm of a trillion galaxies. A trillion is hard to comprehend—

TO A HIGHER POWER

Humans are good at a lot of things, and exceptional at others. One skill in particular that we've evolved to be adept at is *pattern recognition*. We're tremendously skilled at identifying abstract patterns, then using those patterns to draw conclusions, make predictions or classify things. It's an innate ability that we've likely evolved over hundreds of thousands of years to improve our survivability (for example, by recognising animal tracks or identifying safe plants and fruits to eat). It's so entrenched in our brains that we crave patterns and do things subconsciously, which can sometimes be to our detriment. One such blind spot is that we're not good at picturing exponential growth.

Exponential growth is when *something grows* at a rate proportional to how much of that something there is, for example, when something is doubling in size. To highlight this, there's a cute story about an inventor who presented the game of chess to a king. The king loved the game and said he would pay whatever the price for it. The inventor said his price was a simple payment, which he would collect each day: on the first day, one grain of rice on the first square of the chessboard; on the second day, two grains of rice on the second square of the chessboard; on the third day, four grains of rice on the third square; on the fourth day, eight grains of rice on the fourth square, and so on, all the way to the 64th square.

The king agreed, thinking it was a great deal. 'One grain, then two, then four ... this is going to be cheap!'

Unfortunately, exponential growth has a habit of sneaking up on us. A quarter of the way through, on the 16th square, the king needed to pay the inventor 32,768 grains, which is about a litre of rice. Halfway through, on the 32nd square, the king knew he was doomed. He owed more than 2 billion grains of rice (equivalent to about 60 queen-size mattresses stacked on top of each other). Continuing all the way up to the final 64th square, the number owed on this day is 9 billion billion grains of rice—the equivalent of covering the entire country of Australia in a layer of rice 3 cm deep! We don't expect it, but exponential growth happens quickly.

just how much bigger than a billion is it? What about a million? A quick way to gain some context is to figure out what the equivalent number of seconds it's equal to. A million seconds is almost 12 days. A billion seconds is 31.7 *years*. And a trillion seconds is a thousand times that—more than 30,000 years! These numbers are massive. And they only get bigger.

The Milky Way is a fairly typical galaxy, so the number of stars it hosts (we've been using the estimate of 250 billion stars) isn't an unreasonable number to use as an average for all galaxies in the Universe. In the observable Universe, therefore, there are about *100 billion trillion* stars! Okay, so I won't subject you to anymore maths, but I will highlight an important detail about this number that, while a bit obvious, is also really easy to miss: it's about a trillion times bigger than the number of stars in the Milky Way. However many civilisations are in the Milky Way—multiply it by 1,000,000,000,000. That's how many civilisations there could be in the entire Universe.

Again, the immediate rebuttal is that the distances are absurd. The Universe is so big and the distances among galaxies are just too great to be covered. There are two problems with this argument, however. The first is that we're making a big assumption that technology can't advance far enough to tackle travelling eye-watering distances in reasonable timeframes. Perhaps faster-than-light travel remains in the realm of science fiction; it certainly isn't possible with our current understanding of physics. But there may be other ways around it.

Einstein's General Theory of Relativity mathematically supports the idea of 'wormholes'—these are like tunnels that could form between two points in space, sort of like portals, allowing rapid travel between two points, regardless of distance. Mastering the technology that could harness this sort of manipulation of the natural world, warping and moulding spacetime to their every whim, may well be in the sights of a suitably advanced civilisation. The huge distances among galaxies may not even be an issue. Suppose the distances are still an issue though—that wormholes, manipulating spacetime, faster-than-light travel . . . all remain firmly in the world of science fiction. Well, if that's the case then, sure, they may not have visited us, but a trillion galaxies? A hundred billion trillion stars? Ten billion years of existence? The numbers are so great that, even if advanced civilisations were rare, there would have been many that had grown so large, vast and advanced that their existence would be observable from elsewhere in the Universe. Alien megastructures and galactic engineering, artificial or unnatural observations, and technosignatures would be observable over much *greater* distances than a civilisation may be able to physically travel. This argument doesn't require exotic physics outside our current understanding, just the continued and relentless growth of technology to ever greater scales. And so we're back to where we started, faced with Fermi's paradox. The Great Silence.

And that's the boggle of it all. Whether we consider just our galaxy or the entire Universe, the numbers are staggering. It seems almost preposterous that we're alone, and yet here we are. Why? Why don't we see or hear anything? Why do we *appear* to be alone? Is it because we *actually* are alone? Or is it because we're being *left* alone? Of course, no one can answer that question (I certainly won't be able to provide any answers here), but what we can do is try to reason why we've ended up in this particular circumstance. What sorts of things could explain our apparent unique position as the only known life in the Universe?

THE GREAT FILTER

There are a few ways we can slice this problem to try and ascertain the root cause of this paradox; the answers can typically be described referring to something called the *Great Filter*. The Great Filter is some hypothetical 'gate' that all life needs to pass through when going from 'not life' all the way through to an advanced civilisation capable of colonising the entire galaxy. Several steps need to be taken: *abiogenesis* (we discussed this back in Chapter 6—it's the jump from chemical building blocks to the simplest, most basic forms of life); all the critical evolutionary stages; enhanced cognitive abilities; changing from nomadic communities to civilisations; utilising the power of the entire planet; engaging in space exploration; harnessing the power of the entire star system; interplanetary expansion; and so on, all the way through to complete galactic colonisation. There are probably many more steps, but these seem to be some of the most important. So that begs the question: Which one is the Great Filter?

The crux of the problem is that we don't know, and the implications of not knowing are quite frightening. In essence, there are two possibilities: we've passed through the Great Filter already (that is, it's *behind* us), or we haven't passed through the Great Filter yet (that is, it's *in front* of us). Let's tackle the bad news first. If the Great Filter is in front of us, does that mean we're doomed? That's the implication; the reason we haven't heard or seen anyone is that one of those steps we outlined earlier proves far too challenging for life to achieve, and so we're destined to fail as all other life does. It sounds like a bleak outlook.

Looking at the steps, we're at the stage somewhere around 'utilising the power of the entire planet' and 'engaging in space exploration', which means something will stop us somewhere along the next few steps. Interplanetary expansion and harvesting all the power of Sun—these seem like challenges humanity, with all its innovation and ingenuity, could solve. We certainly haven't hit a bottleneck with advancing technology yet, so there's no reason to doubt ourselves and our ability to engineer solutions to problems we face. Perhaps the issue is when we look beyond the Solar System. Maybe it turns out that colonising beyond our own star system is just too hard.

Going through some of the rationale earlier in this chapter, however, that doesn't seem to be the case. Even with the physical limitations we have, on a long-enough timeline, galactic colonisation seems inevitable, so it's likely to be something else.

There's no answer to this question, only hypotheses. To go into them all in just this chapter wouldn't do justice to them, but I will float one here. Perhaps life is often its own undoing. With advanced technology and dreams of expansion, we end up destroying each other, or our planet, either of which certainly prevents us from achieving Milky Way colonisation. It's a grim notion, so let's instead take it and turn it into a kinder and more positive message: we've but one planet to live on, and so we—humanity—need to take care of it and each other to ensure our continued prosperous, harmonious and fulfilled existence. That feels better, right? And if you only take one thing away from reading this book, let it be that.

So if that's what we have to look forward to if the Great Filter is in front of us, then what does it mean if it's behind us? It means that we've succeeded where all other life forms have failed. That we've overcome something incredibly difficult, in which the likelihood of success is tiny. So tiny that it actually impacts some of those enormous numbers we were talking about earlier. It could be a one in 1 billion or 1 in 1 trillion event. That certainly makes a dent in some of our big numbers from earlier. So what could it be?

Again, there's no definitive answer to this, but several steps have been put forward as contenders for what would be considered our greatest triumph. One suggestion is that abiogenesis itself is incredibly rare. That going from a mixture of complex chemical molecules to what would be considered the first form of life is so extraordinarily rare that our mere existence is miraculous. If that's the case, then not only is our Milky Way likely to be devoid of intelligent life, but it's devoid of life *altogether*. If abiogenesis is the Great Filter, this basically asserts that we appear to be alone *because we are*. That's a bit frightening and sad to me, but at least we're not on a doomed timeline with the Great Filter in front of us, so I'm not going to complain if that were indeed the case.

Another suggestion is that one of the early steps in evolution is alarmingly difficult; specifically, the jump from prokaryotic to eukaryotic life. Now, we won't go into detail here (because it's just too deep and complex to cover in a chapter), but the basic premise is that prokaryotes are your first and simplest forms of life. Typically, these are your single-cellular bacteria and microbial life. It's simple, but it's life. Cyanobacteria is a type of prokaryotic life, and, as we mentioned in Chapter 6, they first appeared about 3.5 billion–4 billion years ago. For billions of years, prokaryotes had Earth all to themselves.

Now, eukaryotes are basically all other life. Plants, animals, humans, all complex life falls under the eukaryotes umbrella. The evolutionary step from prokaryotic to eukaryotic life took place somewhere in the realm of 1–2 billion years. This step not only took a long time, but is clearly also crucial to our existence. Without eukaryotes, we wouldn't have nearly the same complexity of life as we know it, and certainly not humans.

Perhaps it's this step that is prodigious. That the Milky Way, and indeed the Universe, is teeming with life but it's prokaryotic (or some equivalent) life: simple life. Life is not rare, but intelligence certainly is.

LEADERS OF THE PACK

Is the Great Filter in front of us or behind us? We just don't know. The first glimpse of alien life may give us enough information to draw a conclusion but, for now, we're constrained to mere speculation. There is also the possibility, however, that there is no Great Filter at all. Perhaps we are just leading the pack. What if the Universe was just a violent and dangerous place for the first 10 billion years or so after the Big Bang? That there was an abundance of chaos and destruction? Collisions and explosions. Bursts and blasts. Highly energetic radiation flying around, preventing any life from surviving, or even starting, in the first place.

It's possible that it just happens to take this long for life to finally have a chance to pop up and begin its evolutionary journey, and we are the first generation. When life began on Earth about 4 billion years ago, that was actually the start of life in the Milky Way. The reason we haven't seen anyone else yet is because we're first; *we're* the technologically advanced civilisation that future intelligent life will search for as evidence that *they're* not alone. That would be quite a humbling position to find ourselves in. Just as we should be caretakers of Earth, we should be custodians of the Milky Way, and care for it and the other life in it. Conservators of the galaxy.

Let me put forward one last consideration for you. One final hypothesis to leave you excited, curious, a little confused but, most importantly, intrigued. What if there's no Great Filter? What if we're not first? What if no one can be seen *because no one wants to be seen?* Intelligent life in the form of advanced civilisations may exist and permeate the entire galaxy—and indeed the Universe—but for some unknown reason they don't want to be seen. It would turn our understanding of ourselves, life and the Universe upside down. Whether it's because of fear, ignorance, carelessness or curiosity, the justification would be profound. We can only guess why it would be the case, but I think that particular discussion is for another book.

In the meantime, we're left with the frustrating reality of Fermi's paradox. Perhaps we will find some clues as to how this mystery unravels within our lifetimes. The answer may just be one detection of life away.

EPILOGUE

WHAT HAVE WE LEARNED?

After all has been said and done, what have we learned? We covered a lot in this book, and it probably goes without saying that this is just the tip of the iceberg. With each chapter, we dived headfirst into something new, and it would be a travesty not to acknowledge how much depth and richness there is for each topic; there are volumes of books and research papers going into detail about them all. There is no way we could possibly cover the breadth of science that has been accomplished with any level of detail in this book. What we did was to weave these fields together to tell the bigger, broader story; we stitched together the scientific observations and findings to depict a grander picture of us, what makes us special and how we belong.

We dissected what it is that makes Earth special. What it is about the planet we call home that has led to it being the only known place where life not only exists, but flourishes. We found there's not one fundamental reason, but a series of serendipitous circumstances that have led to it being the perfect cradle for life. From the unique conditions that allow water to exist as a solid, liquid or gas, to the protective atmosphere and magnetosphere, plate tectonics, the giant planetary guardian that is Jupiter and the abnormally large Moon, the resultant environment has proven to be just right not only for us, but for all life.

We then departed Earth and surveyed our planetary neighbourhood: the Solar System. From our small red neighbour, Mars, to the exotic ocean moons of the gas giants, Jupiter and Saturn, we have found that, throughout the Solar System's existence, there have been times when places other than Earth may have provided conditions conducive to life. Simple life, to be sure, but life no less. Did life exist in some of these worlds in ages past? Does that life still exist? What evidence remains that we can find today? Beyond our Solar System, we cast our net ever wider. Our scientific innovations and ingenuity have enabled us to observe planets orbiting stars other than our Sun. We discussed not only how we look at these planets, but how we obtain information about their properties. How we determine their size, chemical composition, details about their atmosphere and indeed if life exists there.

Finally, we opened our minds wider and considered intelligent civilisations, and the likelihood they could exist. What sort of numbers of civilisations are there likely to be in the Milky Way? What about the whole Universe? We detailed some of the more bizarre observations we've seen in the decades we've spent searching the skies, and discussed how future space missions will continue our search for life and provide us with insight into whether or not we are alone. Whether or not life is commonplace or atypical. We opened up the perplexing conundrum of why we haven't seen anyone yet. Why do we appear to be all alone? Is it because we are alone, or is it because we are being *left* alone?

The take-home message from this book is that, while these big questions can be both scary and exciting, they provide us with important context. They provide us with detail about the nature of physics, life and existence itself. We sit in a privileged position, bearing a responsibility of the highest order. We are the guardians of Earth, the custodians of the planet and, as we oversee the world and animal kingdom, we do so with an extraordinary obligation. Living on the only known planet that bears life, we exist as the only known life form with the cognitive abilities capable of preserving and caring for, not only ourselves but others. We are the only life form who can prepare for the future; the only life form who can ensure the continued survival of ourselves and others. We must not take this moral, social, ethical and existential responsibility lightly. Natural selection played its game and selected us to lead and protect, and we must do so with care, consideration and compassion.

That responsibility is of the utmost importance. It means that whether or not we're alone doesn't change anything about how we should conduct our lives. Whether we exist alone, with the Universe as our private playground, or share it harmoniously coexisting with other life forms, we should always strive to spread love and positivity; treat each other with care, kindness and compassion; and seek to improve and enrich the world, and the lives of the people around us.

ACKNOWLEDGEMENTS

Writing this book has been something I've really wanted to do for some time now, and while it's been a tremendously rewarding and exciting undertaking, it has not been without its challenges. Challenges that I would never have been able to overcome on my own.

First and foremost, I want to thank my mum, Karen, and my dad, Terry, for, well, basically everything, really. Not only have you both been instrumental in helping me to improve, proofread and refine this book to the final state it is in now, but most importantly you have helped me every step of the way throughout my life.

My siblings, thank you for your love and support. My oldest brother, David, thank you for enriching my life with everything you do that is uniquely you (despite your thorough enjoyment of winding me up). I know (as you say) that you always 'have my back'. My older brother, Mark, thank you for the fits of tear-inducing laughter we've enjoyed over the years, and the hilariously dumb stories and personal jokes we've enjoyed that still push our funny buttons to this day. My younger sister, Milly, thank you for lifting me up when I am down. That's no easy feat, but your unique sense of humour always manages to cut through. Thank you for always being there to help me no matter what.

My friends—the family I chose—thank you. There are simply too many of you to name. You have carried me through the bad times, celebrated with me through the good times, and I am so lucky to have you all in my life. This book would never have been started let alone finished without your support.

My academic supervisors, Professor Sarah Maddison, Professor Jonti Horner and Professor Anders Johansen. Thank you for taking the flame of curiosity I had in space, and igniting it into a roaring fire. Your passion for astrophysics is infectious, and your guidance, mentorship and support helped me to nurture that passion for myself, and taught me how to be a scientist.

My management team at Watercooler Talent, David Wilson and Andrew Gaul. Thank you for your support and belief in me. You have helped me navigate a new and unfamiliar world, and have done so with warmth and compassion.

The team at Allen & Unwin that I have worked closely with throughout all of this, especially Annette Barlow and Samantha Kent. Thank you for helping bring this dream to life. Working with you has made this process thoroughly enjoyable.

To everyone else who has helped me on my life journey to this point, thank you. I am grateful for all the help I have been lucky enough to have received, no matter how big or small.

IMAGE SOURCES

Pages i–ii, iii: Daniel Olah/Unsplash

Pages iv–v: NASA/Unsplash

Pages vi–ix: ActionVance/Unsplash

Pages x–1: Daniel Olah/Unsplash

Page 2: NASA/Unsplash

Pages 12–13, lunar eclipse: Zoltan Tasi/Unsplash

Page 14: Shutterstock (background)

Page 20: Shutterstock (background)

Page 22: NASA www.nasa.gov/image-feature/apollo-8-earthrise

Page 33: Shutterstock (top); NASA (bottom) www.nasa.gov/mission_pages/sunearth/news/gallery/20110917-iss-aurora.html

Page 47, the United States from above: NASA/Unsplash

Page 49: NASA/Unsplash

Pages 50–51: Shutterstock (asteroid icons)

Page 55: Shutterstock

Page 63, Jupiter: Planet Volumes/Unsplash

Page 82: Shutterstock

Page 88: Shutterstock (clock face)

Pages 90–91: Vincentiu Solomon/Unsplash

Page 98: Shutterstock

Page 105, Neptune as seen from Voyager 2: NASA/Unsplash

Page 107, the surface of Mars from the Perseverance rover: NASA

Pages 112–13: NASA

Page 124: Shutterstock (Rundle Mall, tennis ball, basketball, dinner plate)

Pages 126–7: NASA https://solarsystem.nasa.gov/missions/cassini/science/enceladus/

Pages 128–9: NASA www.nasa.gov/feature/jpl/enceladus-jets-surprises-in-star light

Page 141, Socorro, New Mexico, USA: Donald Giannatti/Unsplash

Page 153: Shutterstock

Page 160: NASA (left); Shutterstock (right)

Page 165: An astronaut works on the International Space Station, NASA/Unsplash

Page 181: NASA/Unsplash

Pages 182–3, 184–5: Farzad Mohsenvand/Unsplash

Page 185: Juskteez Vu/Unsplash

Pages 194–5: Arthur Yao/Unsplash

Page 197: NASA/Unsplash

Page 215, the Aurora Borealis seen from Iceland: Luke Stackpoole/Unsplash

Page 223: Shutterstock

Page 243: Daniel Olah/Unsplash

All other illustrations and composite images by Simon Rattray, Squirt Creative https://squirtcreative.com

NOTES

1 Walsh, KJ, Morbidelli, A, Raymond, SN, O'Brien, DP & Mandell, AM, 2011, 'A low mass for Mars from Jupiter's early gas-driven migration', *Nature*, vol. 475, pp. 206–9 <https://www.nature.com/articles/nature10201>; and Carroll, M, 2017, *Earths of Distant Suns: How We Find Them, Communicate with Them, and Maybe Even Travel There*, Springer International Publishing, Switzerland <https://link.springer.com/book/10.1007/978-3-319-43964-8>

2 Walsh, KJ, Morbidelli, A, Raymond, SN, O'Brien, DP & Mandell, AM, 2011, 'A low mass for Mars from Jupiter's early gas-driven migration', *Nature*, vol. 475, pp. 206–9 <https://www.nature.com/articles/nature10201>

3 Matson, DL, Spilker, LJ & Lebreton, JP, 2002, 'The Cassini/Huygens Mission to the Saturnian System', *Space Science Reviews*, vol. 104, pp. 1–58 <https://link.springer.com/article/10.1023/A:1023609211620>

4 Exoplanets discovered around a star are named after the star they orbit and then are designated a letter from the alphabet according to the order in which they are discovered. Strangely, the first exoplanet

discovered is given the designation 'b', for example, 51 Peg b. The reason we start at 'b' is because the star of a planetary system is considered to be the 'a'—the first object discovered in the system.

5 Simpson, F, 2016, 'The size distribution of inhabited planets', *Monthly Notices of the Royal Astronomical Society: Letters*, vol. 456, no. 1, pp. L59–L63 <https://academic.oup.com/mnrasl/article/456/1/L59/2589573>

INDEX

First published in 2022
Copyright © 2022 Matt Agnew

All rights reserved. No part of this book may be reproduced or transmitted in any form or by any means, electronic or mechanical, including photocopying, recording or by any information storage and retrieval system, without prior permission in writing from the publisher. The Australian Copyright Act 1968 (the Act) allows a maximum of one chapter or 10 per cent of this book, whichever is the greater, to be photocopied by any educational institution for its educational purposes provided that the educational institution (or body that administers it) has given a remuneration notice to the Copyright Agency (Australia) under the Act.

Allen & Unwin
Cammeraygal Country
83 Alexander Street
Crows Nest NSW 2065
Australia
Phone: (61 2) 8425 0100
Email: info@allenandunwin.com
Web: www.allenandunwin.com

Allen & Unwin acknowledges the Traditional Owners of the Country on which we live and work. We pay our respects to all Aboriginal and Torres Strait Islander Elders, past and present.

A catalogue record for this book is available from the National Library of Australia

ISBN 978 1 76106 518 7

Illustrations by Simon Rattray, Squirt Creative
Index by Puddingburn
Internal design by Louisa Maggio Design
Set in 12 pt Gill Sans by Louisa Maggio Design
Printed and bound in China by C&C Offset Printing Co., Ltd.

10 9 8 7 6 5 4 3 2 1